North Sea Climate

based on observations from ships and lightvessels

North Sea Climate

based on observations
from ships and lightvessels

by

C. G. KOREVAAR

Royal Dutch Meteorological Institute (KNMI),
De Bilt, The Netherlands

KLUWER ACADEMIC PUBLISHERS
DORDRECHT / BOSTON / LONDON

Library of Congress Cataloging in Publication Data

Korevaar, C. G.
    North Sea climate : based on observations from ships and
  lightvessels / C.G. Korevaar.
        p.   cm.
    ISBN 0-7923-0664-3 (alk. paper)
    1. North Sea Region--Climate--Charts, diagrams, etc.
  2. Meteorology, Maritime--North Sea Region--Charts, diagrams, etc.
  I. Title.
  QC994.2.K68   1990
  551.69163'36--dc20
                                                    90-34914
                                                         CIP

ISBN 0-7923-0664-3

Published by Kluwer Academic Publishers,
P.O. Box 17, 3300 AA Dordrecht, The Netherlands.

Kluwer Academic Publishers incorporates
the publishing programmes of
D. Reidel, Martinus Nijhoff, Dr W. Junk and MTP Press.

Sold and distributed in the U.S.A. and Canada
by Kluwer Academic Publishers,
101 Philip Drive, Norwell, MA 02061, U.S.A.

In all other countries, sold and distributed
by Kluwer Academic Publishers Group,
P.O. Box 322, 3300 AH Dordrecht, The Netherlands.

LIST OF CONTENTS

INDEX TO FIGURES

Figure captions

Figure 1 Selected areas and lightvessels for which statistical data are given in figures 10-16, 25-32, 45-61, 65, 66, 79-96.

Figures 2-5 Mean air temperatures with standard deviations for the months February, May, August and November over the period 1961-1980.

Figures 6-9 Mean sea surface temperatures with standard deviations for the months February, May, August and November over the period 1961-1980.

Figure 10 Annual variation of the air temperature (Ta) and sea surface temperature (Ts) for a number of subareas and lightvessels indicated in figure 1.

Figure 11 Frequency distribution of the air temperature with mean and standard deviation per month and for all months together for lightvessel Texel.

Figure 12 Frequency distribution of the sea surface temperature with mean and standard deviation per month and for all months together for lightvessel Texel.

Figure 13 Variation of mean air temperature (winter and summer months) at Noord Hinder (1859-1980).

Figure 14 Variation of mean air temperature (winter and summer months) at Haaks/Texel (1890-1977).

Figure 15 Variation of mean sea surface temperature (winter and summer months) at Noord Hinder (1885-1980).

Figure 16 Variation of mean sea surface temperature (winter and summer months) at Haaks/Texel (1890-1977).

Figures 17-24 Frequencies of visibility less than 4000 m and 1000 m for the months January, April, July and October over the period 1961-1980.

Figure 25 Annual variation of the frequency of occurrence of visibilities (V) < 1 km and ≥ 10 km, cloud cover (N) ≤ 2/8 and ≥ 6/8 and of precipitation (R) for a number of subareas and lightvessels indicated in figure 1.

Figures 26-29 Seasonal persistence diagrams for visibilities ≤ 4000 m, ≤ 1000 m and ≤ 200 m at lightvessel Texel.

Figure 30 Annual variation of sea level pressure (P) for thesubareas indicated in figure 1.

Figures 31,32 Prevailing wind directions and steadiness of the wind for areas indicated in figure 1.

Figures 33-44 Frequencies of wind forces greater than 5 and 7 Beaufort and less than 4 Beaufort for the months January, April, July and October over the period 1961-1980.

Figure 45 Annual variation of the frequency of occurrence of wind force ≥ 8, ≥ 6 and ≤ 2 for a number of subareas and lightvessels indicated in figure 1.

Figures 46-57 Wind roses for three selected areas (areas 03, 15 and 24 in figure 1) for the months January, April, July and October over the period 1961-1980.

Figures 58-61 Persistence diagrams for winter and summer season giving the durations of periods with wind forces ≥ 6, 7, 8, 9 and 10 Beaufort and ≤ 2, 3, 4 and 5 Beaufort at lightvessel Texel.

Figures 62-64 Extreme hourly mean wind speeds in m/s with return periods of 10, 50 and 100 years.

Figure 65 Comparison of frequencies of wind forces ≥ 4, ≥ 6, ≥ 8 and ≥ 10 Beaufort at Netherlands lightvessels in different time periods.

Figure 66 Yearly numbers of storms (wind force ≥ 10) at lightvessel Texel (L.V. Haaks until 1939) for the years 1891-1980.

Figures 67-78 Frequencies of wave heights greater than 1.75 and 3.76 m and less than 1.25 m for the months January, April, July and October over the period 1961-1980.

Figure 79 Annual variation of the frequency of occurrence of wave heights ≥ 4 m, ≥ 6 m, ≤ 1.5 m and of wave periods ≥ 6 seconds for a number of subareas and lightvessels indicated in figure 1.

Figures 80-91 Wave roses for the three selected areas (areas 03, 15 and 24 in figure 1) for the months January, April, July and October over the period 1961-1980.

Figure 92 Prevailing swell directions for areas indicated in figure 1.

Figures 93-96 Persistence diagrams for winter and summer season giving the durations of periods with wave heights > 0.75, 1.75, 2.75 and 4.75 m and < 0.25, 0.75, 1.25, 1.75 and 2.25 m at lightvessel Texel.

Figures 97-99 Extreme wave heights (based on visual observations) in metres with return periods of 10, 50 and 100 years.

PREFACE

The need for climatological data of the North Sea has increased during the past years. The increase in offshore and recreational activities can benefit greatly from such data. In order to meet this need weather observations made by ships and lightvessels in the North Sea have been processed extensively, which has resulted in a large number of tables, charts, etc. with climatological data.

This publication gives a selection of these data, in which the emphasis lies on wind and wave data. In addition, some characteristic data on air and sea temperature, cloud cover, precipitation, visibility and sea level pressure are given.

With regard to the observations of the lightvessels the publication can be considered as a continuation of earlier work. It also concludes the era of observations made by crew on board lightvessels, which gradually ended in the 1970's. The observations on fixed platforms which are now replacing the observations of the lightvessels are of a quite different character.

The data are based on observations made on board lightvessels during the period 1949 - 1980 and voluntary observing ships in the area between 51°N and 60°N during the period 1961-1980. The observations of the lightvessels have been compared with those published for earlier periods (1859 - 1883, 1884 - 1909 and 1910 - 1940).

The manuscript (or part of it) has been critically examined by Prof.Dr.Ir. J.A. Battjes, E. Bouws, Dr.Ir. J.A. Buishand, H.A.M. Geurts, Dr. G.J.Komen, Dr. G.P. Können, R.A. van Moerkerken and Prof.Dr.Ir. H. Tennekes. I am indebted to them for their helpful suggestions. Special acknowledgement is given to R.R. Broersma and J.M. Koopsta who performed the data processing and to J.W. Schaap and Ir. M.P. Visser who coordinated the work to be done for the publication. I want to express my greatest appreciation to Mrs. G. de Graaf-Boshuis and Mrs. M. van Lahr who typed the manuscript and to M. Latupeirissa and R.J. Meijer of the KNMI Studio who made the text print-ready and took care of the drawings.

C.G. Korevaar

Chapter 1
Introduction

Due to the increasing activities on the North Sea, KNMI receives many questions concerning the climate of this sea. In order to be able to answer these questions the weather observations made by selected ships and lightvessels in the North Sea have been processed extensively during the past years. This resulted in a large number of tables, diagrams, graphs, roses and charts with climatological data. Part of this data has been published in two separate scientific reports (Korevaar 1987, 1989).

The purpose of this publication is to present a selection of the collected data to the users - in the fields of the offshore industry, shipping, fishing, dredging, port- and harbour construction, coastal defense or recreation - in order to give an answer to the most general questions about the climatological conditions in the North Sea. In case information is needed in more detail, KNMI can always be consulted. At present also data from fixed platforms and buoys are collected.

The emphasis lies on wind and wave data, because practice learned that the greater part of the questions received are related to these data, firstly for selecting workable periods and secondly for determining design criteria of maritime structures. In addition to wind and wave data also some characteristic data on air and sea temperature, cloud cover, precipitation, visibility and sea level pressure are given.

Both the lightvessels and the selected ships belong to the global observing net-work of the World Meteorological Organization (WMO). In the first place the observations are used to make synoptic weather charts, but by storing them they can also be used for climatological purposes. The observations of the lightvessels are taken at fixed positions. This has the advantage that at these positions long time series have been created by which it is also possible to determine persis-tencies of certain conditions. All lightvessels are situated near the coast. For infor-mation about the central and northern parts of the North Sea the observations of the selected ships - mostly merchant ships plowing from port to port - are more suitable. In contrast with those of the lightvessels the observations on board selected ships are taken at variable positions. However, observational methods are standard for all ships involved. Both the wind and wave data are mostly based on visual estimates.

The number of Netherlands selected ships observations kept in the KNMI data base was too small to give a homogeneous distribution in space and time. There-fore use has been made of the WMO system for collection of ship- weather-reports of maritime nations. According to this system, also known as the Marine Climatological Summaries Scheme, a number of countries has the responsibility for collecting ships- observations since the year 1961. One of these countries is the Netherlands, which is responsible for collecting ships-observations in the Mediterranean and the southern part of the Indian Ocean, and Great Britain for the eastern part of the North Atlantic, including the North Sea. By mutual exchange with Great Britain about 700.000 ships-observations for the North Sea area between 51°N and 60°N were obtained over the period 1961-1980. With these observations a reasonably good coverage in space and time has been obtained with an average number of at least one observation per day for a great number of one-degree-squares (squares of one degree latitude by one degree longitude) and for most other one-degree-squares an average of at least one observation per two days.

The observations are stored on magnetic tape. They have been subject to a quality control programme. Among other things, the data in one observation should be internally consistent. For example, it is not allowed that at low wind-speed a high sea is reported or that the present weather code indicates fog, while the visibility code gives good visibility. In addition, data has been checked if certain physical and climatological limits have not been exceeded.

According to the definition of the WMO a selected ship station is a mobile ship-station that is equipped with sufficient certified meteorological instruments for making observations and that transmits required observations in the appropriate code for ships. The observations of the selected ships are mostly made at six-hour intervals (standard times are 00, 06, 12 and 18 hours GMT). Most observations are found along the main shipping routes. On board the lightvessels the observations are generally made at three-hour intervals. Although a crew are in general not professional observers, at nautical schools they receive training in meteorology and by making the observations they build up experience.

While for the processing of the observations of the selected ships the period 1961-1980 was most appropriate, for the lightvessels the period of ca. 1950 to ca. 1980 has been used mainly. In the past, reports of the meteorological and some oceanographic observations made on board the Netherlands lightvessels have been published by J.P. v.d. Stok (1912) for the period 1859-1910, by G. Verploegh (1956-1959) for the period 1910-1939 and by C.G. Korevaar (1987) for the period 1949-1980. Some data on the wind and wave frequencies were given by R. Dorrestein (1967), mainly for the period 1949-1957 and by E. Bouws (1978) for the period 1949-1975.

Chapter 2
Methods of observation and reporting

The observations are part of the global observing network data of the WMO. They mainly serve for the preparation of weather charts. Apart from this, they are collected, quality controlled and stored for climatological purposes. Some elements are observed with instruments, such as air pressure and air and water temperature. Some phenomena such as rain, snow, thunderstorms, cloud types are simply determined. Finally there are a number of elements which must be estimated from visual observations. To this category belong cloud cover, visibility, wind and waves.

In the following, the method of observation of the most important elements will be described in brief.

2.1 Air temperature (and humidity)
Dry bulb temperature (air temperature) and wet bulb temperature are measured with a sling psychrometer or an aspirated psychrometer which are preferably exposed on the windward side of the bridge in a stream of air, fresh from the sea, which has not been in contact with, or passed over, the ship and is adequately shielded from radiation, precipitation and spray. The air temperature is reported in tenths of degrees.

2.2 Sea surface temperature
The temperature of the water in the near surface layer may be observed by several methods. Most common methods are: taking a sample with a specially designed sea-bucket or reading the temperature of the condenser intake water. Some ships have electrical thermometers which measure either directly or through the hull. Comparative investigations of the various techniques are collected by Terziev (1981). The sea surface temperature is reported in tenths of degrees Celsius.

2.3 Air pressure
The air pressure is measured either by a precision aneroid or by a mercury barometer. After reducing the measurements to sea level they are reported in tenths of hectopascals.

2.4 Cloud cover
The total amount of cloud is estimated by considering how much of the sky is covered by cloud and is reported in oktas (one okta means one-eighth of the sky dome as seen by the observer). The reporting code is as follows.

0 = none
1 = 1 okta or less, but not zero
2 = 2 oktas
3 = 3 oktas
4 = 4 oktas

5 = 5 oktas
6 = 6 oktas
7 = 7 oktas
8 = 8 oktas
9 = sky obscured or cloud amount cannot
       be estimated.

2.5 Precipitation

The occurrence of precipitation is simply determined and reported by the present weather (ww) code (WMO code 4677). By this code one hundred different weather situations at the time of observation can be reported. The first figure of the scale ww indicates a division of the scale in ten deciles, which corresponds to ten principal categories of weather at the time of observation; e.g. with ww = 50-59 various kinds of drizzle can be indicated, while ww = 60-69 means rain, ww = 70-79: solid precipitation not in shower, ww = 80-99: showery precipitation. Measuring the amount of precipitation on a sailing ship is hardly possible. Therefore this is only measured on board the lightvessels by means of a conical rain-gauge attached to the stay between the foremast and the lighttower.

2.6 Visibility

Visibility describes the degree of transparency of the atmosphere and is defined as the maximum distance at which an object can be seen, for example: in normal daylight, the distance at which an object such as a ship can just be discerned as such by an observer with normal eyes and in darkness, the distance at which a light of certain intensity and colour is just visible.

On board ships the horizontal visibility is almost always estimated. At sea a detailed determination of visibility is very difficult. For this reason a rather coarse scale is used for the reporting code (VV), which is as follows:

90	< 50 m	
91	50 - 200 m	
92	200 - 500 m	
93	500 - 1000 m	
94	1 - 2 km	
95	2 - 4 km	
96	4 - 10 km	
97	10 - 20 km	
98	20 - 50 km	
99	> 50 km	

If the horizontal visibility is not the same in different directions with VV the smallest visibility is reported.

2.7 Wind

Most wind observations are estimates. In the period 1961-1970 only about 10% were measurements. In the decade 1971-1980 the number of measurements has increased considerably in some areas, often to about 30% and sometimes even more. In both decades in the German Bight about 70% of the observations consisted of measurements.

In the case of estimates, the direction from which the wind blows, is determined with the compass from the direction of the crests of the sea waves (the wind direction is perpendicular to this) or from the direction of the foam streaks, which are blown in the direction of the wind. The wind direction is reported in tens of degrees.

The wind force is estimated using the Beaufort Scale. According to this scale the wind force can be given in 13 numbers, from 0 to 12. The number 0 corresponds to calm (no wind at all), while the number 12 corresponds to hurricane force. This scale has been developed by the British admiral Sir Francis Beaufort in 1805 and is based on the speed a fullrigged frigate could make for the lower wind speeds and on the quantity of sails the ship could carry for the higher wind speeds. Later when the time of sailing vessels was over, the German captain P. Petersen has added to each scale number a description of the visible effect of wind force on the sea surface. In this way it became possible to estimate the wind force from the appearance of the sea. Since 1946 there is an official WMO conversion scale in which to each scale number an interval of equivalent wind speeds (in knots and for a height of 10 meters above the sea surface) has been assigned. This scale is given in appendix 1.

Although it has appeared from several studies (Verploegh 1956, Dury 1970, Kaufeld 1981) that in reality the relation between Beaufort numbers and wind speed deviates considerably from the official WMO conversion table all efforts in the WMO Commission for Marine Meteorology to change it have been unsuccessful. Only for scientific use, other equivalent windspeeds have been accepted. This so-called scientific scale can be found in appendix 2.

To avoid ambiguity, in this study mainly Beaufort numbers have been used, in agreement with the observational data.

The accuracy of the estimates is influenced among others by the fact that the appearance of the sea is not only determined by the wind. Other phenomena which play a role are: the influence of the plankton content of sea water on the formation of foam; the influence of the current and bottom depth on the form of the waves; the influence of the air stability on the steepness of the waves; the influence of heavy rainfall or oil pollution; the influence of the wind fetch. Nevertheless the accuracy with which wind-speed can be estimated is still reasonable.

Verploegh (1967) found for the standard deviation 0.58 I (I being the width of the scale interval) for each of the steps 1 to 10 of the Beaufort scale. This means that the standard deviation of an individual wind speed observation varies from 0.76 metres per second at step one (mean wind speed of 2.0 m/s) to 1.34 metres per second at step five (mean wind speed of 10.2 m/s) and 2.6 metres per second at step ten (mean wind speed of 24.2 m/s).

The averaging or representative time for these observations is not really known. In a study by Graham (1982) in which the winds estimated on board voluntary observing ships are compared with instrumental measurements af fixed positions, the averaging time has been taken as equivalent to an hour. Owing to the relatively slow response of the sea to changes in wind speed this seems reasonable.

Winds measured by means of an anemometer on board ship give an accurate representation of the speed and direction of the air flow over the ship at the location of the anemometer. In order to determine the actual wind over the sea, the measured wind should be corrected to allow for the ship's movement by means of a vector diagram. The main problem with measuring wind on board a ship or a platform is that a fixed anemometer often cannot be exposed sufficiently to all wind directions; consequently the readings from it may not be representative of the undisturbed air flow. Errors in the derived true wind thus occur, which may be appreciable, particularly with following winds. On big ships and on oil rigs, anemometers are usually installed at great heights; heights of 40 metres above the surface of the sea are by no means uncommon. The wind speed normally increases with height, the height variation depending on the air stability. Routine observations are, however, not corrected for heights. This is yet another source of error. Errors of measured winds in ship reports can be so large that one should not automatically accept such a report as being accurate. When errors occur, they are often of the same order of magnitude as those in a visual observation; errors of about two metres per second may occur at all wind speeds.

2.8 Waves

Ships observations of waves are done visually. The parametres which are estimated are direction, period and height. The direction of the sea (or windwaves: waves clearly related to the prevailing wind) is taken to be equal to the direction of the wind. If a separate swell (waves outside the wind field by which they were generated) can be distinguished this is also reported. The wave height is the vertical distance between trough and crest, while the wave period is the time between the passage of two successive wave crests past a fixed point. The mean direction from which the waves are coming is reported in tens of degrees with respect to true North and can be estimated easily by means of the compass since the direction is perpendicular to the wave crests.

The average period is determined by counting the seconds between the passing of a number of well-developed wave crests. Until 1 January 1968 the period had been reported as a code figure P_w (WMO code 1955) as follows.

2 = 5 seconds or less   7 = 14 or 15 seconds
3 = 6 or 7 seconds   8 = 16 or 17 seconds
4 = 8 or 9 seconds   9 = 18 or 19 seconds
5 = 10 or 11 seconds   0 = 20 or 21 seconds
6 = 12 or 13 seconds   1 = 22 seconds or more
/ = calm or period not determined

From 1 January 1968 the meteorological code for ship's observations was changed. Starting from that date the period of the sea has been reported in seconds and the period of the swell according to WMO code 3155 which is as follows.

0 = 10 seconds   5 = 5 seconds or less
1 = 11 seconds   6 = 6 seconds
2 = 12 seconds   7 = 7 seconds
3 = 13 seconds   8 = 8 seconds
4 = 14 seconds or more   9 = 9 seconds
/ = calm or period not determined

For the reporting of the wave height an estimate in half metres is made of the average height of the higher, well-developed waves in the central part of wave groups. Several studies have shown that this height is a fair approximation of the so-called significant wave height, which is defined as the average height of the highest one third part of the waves in the system (cf. section 3.9.5 and Norden-strøm, 1969, Laing,1985).

With respect to the accuracy of visual wave height observations Verploegh (1961) found that the standard error of an individual observation varies from 0.3 m at 1.5 m wave height to one metre at a 6 m wave height.

The visually estimated wave period is - unlike what applies to the wave height - not equal to the significant period (the mean period of the highest one third part of the waves) but is generally smaller. This is due to the bias of a visual observation towards those waves which are steepest.

Chapter 3
Climatological data

In order to give as much information as possible for the area of the North Sea between 51°N and 60°N and for the different months of the year without making this publication too voluminous, the following compromise has been found. Only for 4 months charts are presented with data for the whole area. Mostly January, April, July and October, being the central months of the successive seasons, have been chosen. However, for the temperatures February, May, August and November have been chosen, because at sea the lowest temperatures occur in February and the highest in August. To get an idea about what is happening in the other months next to these charts graphs are given for a number of selected areas and lightvessels (see figure 1) with the annual variation, based on the successive monthly means or frequencies.

3.1 Air temperature

3.1.1 General remarks
The temperature of the air in the lowest layer of the atmosphere above the sea is mainly determined by the surface temperature of the water underneath. Owing to the current system in the North Atlantic Ocean with branches of the so-called North Atlantic current (which is the continuation of the warm Gulf Stream) through the English Channel and around the north of Scotland the mean temperatures over the North Sea are above the average compared with other sea areas at the same geographical latitudes elsewhere.

3.1.2 Geographical distribution
Figures 2 - 5 show the mean air temperatures together with their standard deviations for one-degree-squares (squares of one degree latitude by one-degree-longitude) for the months February, May, August and November over the period 1961-1980. In the geographical distribution of the temperatures the influence of the ocean in the north-west and of the continent in the east can be seen. For instance, the mean temperatures in February are highest in the western areas with values of about 5 degrees and lowest in the German Bight with values of about 1 degree, while to the contrary in August the lowest mean temperatures occur in the north-west (about 13°) and the highest mean temperatures in the south-east (about 17°). The greater range in the south-east part of the region reflects the climate influence of the continent. While the isotherms in February and August have a more north-south orientation, in the months May and November they tend to follow the parallels of latitude. When taking the average of the monthly minimum and maximum temperatures it appears that in practically the whole North Sea these are respectively roughly 3-4 degrees lower and higher than the mean temperatures.

3.1.3 Annual variation
In figure 10 for a number of subareas and lightvessels (shown in figure 1) graphs are given with the annual variation of the air temperature, based on the successive monthly means. Also the annual means have been indicated. The mean annual temperatures do not differ very much from place to place. In the northern areas

they lie between 8 and 10 degrees centigrade and in the southern areas between 9 and 11 degrees. The month with the highest mean temperature is August, while the lowest mean temperature occurs in February, which is a month later than on the continent. The annual variation is largest in the eastern areas which are closest to the continent.

3.1.4 Variability at a certain place
The variability of the temperature at a certain place is shown in the table of figure 11, in which a frequency distribution has been given for the lightvessel Texel over the years 1949-1977. In this table the observations have been arranged into intervals of one degree Celsius. The table shows that the variability is greater in the winter months than in the summer months. The standard deviations for the temperature values given are also representative for other areas near the Dutch, German and Danish coasts. More in the open sea and near the British coasts the variability is somewhat smaller.

3.1.5 Variation over the period 1859-1980
The air temperature does not only vary from place to place or from month to month, but also from year to year and from period to period. An example has been given in figure 13, in which the variations of the mean air temperatures for lightvessel Noord Hinder during the winter minimum and the summer maximum are given. As the annual minimum temperature may fall either in January or in February the mean temperature of these two months have been given in the figure. For the maximum the mean value of July and August has been taken. Noord Hinder has been chosen, because from this lightvessel the longest time series at about the same position is available, namely from 1859-1980, alas with the exception of some years during the first World War and 14 years (1939-1953) during and around the second World War. The figure shows that there can be considerable differences from year to year. Looking at the whole period 1859-1980, no periodicities or obvious trends can be discovered, but there are still some remarkable features. The first period lasting from 1859 to 1885 is apparently warmer than the rest of the time. This is most pronounced in the winter temperatures, which during this period are about one degree Celsius higher than the average over the whole period. Also the summer temperatures are somewhat higher in this period. After 1885 this warm period ended rather abruptly. From 1886 to 1897 the mean winter temperature fluctuated around a much lower level. After 1897 no pronounced cold or warm periods have occurred in the winter temperature. What did occur, were a few abnormally cold extremes, the first in 1929 with a mean winter temperature (January, February) of 2.1°C. The coldest winter in the record was that of the year 1963 with a mean temperature of 0°C. Also 1979 had a cold winter with a mean temperature of 2.8°C. The summer temperatures do not show such large deviations. Here, however, the whole period 1953-1980 is characterised by a low mean summer temperature of about 0.6°C below the overal average. About the same picture is found at the more northerly situated lightvessels, for example at Haaks/Texel (figure 14). However, here the data

started not until 1890. For this reason it cannot be verified whether in the time prior to that year there was also a warmer period here. The resemblance in the rest of the temperature series at these vessels suggests that this supposition may be correct. A clear difference can be seen in the summer temperatures in the period after the second World War. At Noord Hinder they were relatively low, while at Texel they were close to the average.

3.2 Sea surface temperature

3.2.1 Geographical distribution
Figures 6-9 show the mean sea surface temperatures for one degree squares for the months February, May, August and November over the period 1961-1980. Due to the supply of warmer water through the English Channel and around the north of Scotland the mean temperatures in February are highest with values of about 7°C in a tongue of warmer water in the central part of the southern North Sea and in the north-western part of the North Sea. As for the air temperature the lowest mean sea surface temperature (2-3°C) occurs in the German Bight. August gives a same pattern as for the air temperature with the lowest mean temperature (about 12.5°C) in the north-west and the highest mean temperature (about 17.5°C) in the south-east along the Danish, German and Dutch coasts. May and November are transition months in which the mean sea surface temperature in the North Sea does not vary very much from place to place. The mean monthly minimum and maximum temperatures differ roughly 2-3 degrees with the mean temperatures in practically the whole North Sea.

3.2.2 Annual variation
Next to the annual variation of the air temperature in figure 10 also the annual variation as well as the annual mean of the sea surface temperature is given for the same subareas and lightvessels (see figure 1). The mean annual sea surface temperature lies between 9 an 10°C in the northern areas and between 10 and 12° in the southern areas. The month with the highest mean temperature is - like for the air temperature - August, while the lowest mean sea surface temperature mostly occurs in March (which was February for the air temperature). The sea surface temperature shows less variation (smaller standard deviations) than the air temperature. The variations are smallest in the north-west, while the largest variations occur in the German Bight.

3.2.3 Variability at a certain place
The variability of the sea surface temperature at a certain place is shown in the table of figure 12, in which a frequency distribution has been given for the lightvessel Texel over the years 1949-1977. As for the air temperature (in figure 11) the observations have been arranged into intervals of one degree Celsius. The variability is, especially in the winter months, much smaller than that of the the air temperature, while also here the variability is smaller in the summer months than in the winter months.

3.2.4 Variation over the period 1885-1980

The variations of the mean sea surface temperature during the winter months January and February and the summer months July and August for the light-vessels Noord Hinder and Haaks/Texel have been given in figures 15 and 16. For the first period from 1859 to 1884 no temperatures of the sea water are available. For the rest of the time the variations in the sea surface temperature from year to year are similar to these of the air temperature. The large cold peaks in air temperature are not always found back in the water temperature. The cold peak of 1963, however, is clearly very pronounced in the variation of the water temperature.

3.3 Air-sea temperature difference

From the graphs in figure 10, where the mean monthly and annual air and sea surface temperatures are given, it can be seen that in the area of the North Sea the mean air temperature is lower than the sea surface temperature. Only from about April to about July the sea is slightly colder than the overlying air, which is one of the conditions that are favourable for the development of sea fog. These stable conditions also cause relatively low wind speeds during these months.

3.4 Visibility

3.4.1 Geographical distribution

Figures 17-24 give frequencies of visibilities less than 4000 m and 1000 m for one-degree squares for the months January, April, July and October over the period 1961-1980. By definition the visibility in fog is less than 1000 m. Generally the chances of fog are not very high in the North Sea, but there are remarkable differences, both from place to place and from season to season. For example in January practically no fog occurs in the north-western part of the North Sea, while in the south-eastern part the frequency is higher than 5%, in the German Bight even higher than 10%. This situation reverses in July.

3.4.2 Annual variation

In figure 25, graphs are presented with annual variations of the frequency of occurrence of visibilities less than 1 km and equal or more than 10 km for a number of subareas and lightvessels (shown in figure 1). Also the annual means have been indicated. The mean annual percentages for visibilities less than 1 km (such as in fog) are rather small (2-4%). The months of maximum fog are April, May and June (in these months the sea is on the average slightly colder than the overlying air) in the northern areas and January to June in the southern areas when about 5-10% of the observations have a visibility of less than 1 km. During the autumn months the frequency of fog is generally low due to the relatively warm water with respect to the temperature of the air. The frequencies of observations with a visibility of 10 km or more are greater in the last six months of the year than in the first six months.

The following facts concerning the occurrence of fog have been gathered from a study of the Netherlands lightvessels only (Korevaar, 1987).

18

3.4.3 Diurnal variation
A diurnal variation hardly exists. It may be said that most of the fog occurs in the morning hours between 6 and 12 o'clock and the least fog between 18 and 24 o'clock. However, there is no clear distinction.

3.4.4 Relation to wind force
The chance of fog is strongly related to the wind force. For example at lightvessel Goeree (position 51°55'N, 3°39'E) in January the average chance of fog is 4.7%, however, at wind force 0 (calm): 28%, at wind force 1: 19%, at wind force 2: 10%, at wind force 3: 8%, at wind force 4: 3.5%, at wind force 5: 1.5% and at wind force 6 and 7: 0.2%. Generally it can be concluded that the chance of fog decreases with an increase of the wind force and lies above the average at forces 0-3 and below the average at greater wind forces. At wind force 6 or more practically no fog occurs.

3.4.5 Relation to wind direction
The occurrence of fog is also dependent on the wind direction. In the month January at lightvessel Goeree at wind directions between SE and SW the percentages of observations with fog vary from 6-10% (with a general chance of fog of 4.7%), while at wind from the North practically no fog occurs. It appears that during the winter half year fog by preference occurs at calm weather or at winds from directions between SW and East. In the summer half year, when the frequency of fog-occurrence is low, just northerly wind directions are slightly favourite.

3.4.6 Persistence
For the Netherlands lightvessels the duration (persistence) of periods with certain visibility conditions (≤ 4000 m, ≤ 1000 m and ≤ 200 m) have been studied. In figures 26-29 persistence diagrams are given for lightvessel Texel (position 53° 01'N, 4°22'E) for each of the four seasons. It appears that the number of periods with a decreased visibility are larger during winter and spring and smaller in summer and autumn. Only little have a duration of 24 hours or more. At lightvessel Texel the percentage of periods with fog (visibility ≤ 1000 m) and a duration of 24 hours or more is 6.7% in winter, 5.0% in spring, 1.6% in summer and 2.9% in autumn. The decrease of the number of periods with the increase of the duration is rather fast. For the shorter durations to 9-12 hours the decrease is practically exponential. For longer durations the decrease is somewhat slower. The mean durations decrease with decreasing visibility.

3.5 Cloud cover
In general, cloud amounts are high over the North Sea. Figure 25 shows for selected areas (see figure 1) the monthly and annual percentages of surface observations having a cloud cover of 2/8 or less and 6/8 or more. The mean annual percentages are between 11 and 21% for a cloud cover of 2/8 or less and between 60 and 71% for a cloud cover of 6/8 or more. The cloud cover varies little form place to place. Generally, the cloud cover decreases a little from north to south

and from west to east. The total cloud cover also varies with the seasons. The percentages with a cloud cover of 2/8 or less are highest (about 20-30%) during the months April, May and June and lowest (about 10-20%) during the months November, December and January, while, on the contrary, percentages with a cloud cover of 6/8 or more are highest (70-80%) in December, January and February and lowest (50-60%) in April, May and June. It is remarkable that during the summer season there often is a secondary maximum in cloud cover in July.

3.6 Precipitation
In figure 25 the mean monthly and annual percentages of observations with precipitation are given for selected subareas (figure 1). The mean annual percentages are between 6 and 14%. They are higher in the north than in the south. There are no distinct wet or dry seasons, but there is a well-marked seasonal variation. May, June and July are the driest (percentages of about 4-11) and November, December and January are the wettest (percentages of about 11-23). Most of the precipitation falls in the form of rain or drizzle, but especially in January and February an important part is formed by snow. The amount of precipitation is only measured at some lightvessels. The table below gives mean precipitation amounts at the lightvessels Texel (position 53°01'N, 4°22'E) and Noord Hinder (51°39'N, 2°34'E) over a period of 10 years (1968-1977). The amounts are highest in autumn and lowest in spring.

Light-vessel	Precipitation amount (in mm), averaged over the years 1968-1977												
	J	F	M	A	M	J	J	A	S	O	N	D	Year
Texel	60.3	45.2	36.3	33.6	34.3	33.1	53.0	47.3	90.2	66.4	99.0	59.4	658.1
Noord Hinder	44.7	33.4	39.0	29.6	33.4	44.4	37.3	44.6	73.8	67.7	68.3	31.3	536.4

3.7 Sea level pressure
In agreement with the general pressure distribution with an Icelandic Low and an Azores High the mean annual pressure increases from north to south from about 1011 hPa to about 1015 hPa. In figure 30 the annual variation and mean are given for 8 selected subareas (see figure 1). The mean pressure is generally higher in the summer months and lower in the winter months, but there is no smooth seasonal variation. In some areas also in February the pressure is above the average. The lowest mean pressure mostly occurs in November, which is also one of the windiest months.

3.8 Wind

3.8.1 General remarks
The North Sea lies in the zone between the Icelandic Low and the Azores High in which normally depressions with their frontal systems are moving eastwards. Only sometimes a blocking high occurs in or near the region. Therefore winds from westerly directions are prevailing, although winds from all directions are possible. In figure 32 the prevailing wind direction is given by the resultant vector mean direction with each speed set equal to 1, and the steadiness (or constancy) by the ratio of the mean resultant wind speed to the mean wind speed irrespective of direction; in other words the ratio of the vector mean to the scalar mean. In regions where the wind always has about the same direction the ratio will be near 1. For example in the trade winds steadinesses of more than 0.8 occur. The smaller the ratio the more variable the winds are. In figure 32 prevailing monthly and annual wind directions are shown together with the steadiness of the wind for a number of selected areas (figure 1) for the two decades 1961-1970 and 1971-1980. As figure 32 shows the prevailing annual wind direction lies between SW and WNW with a steadiness of 0.2. However, there is a seasonal variation. During the summer months, especially June and July the prevailing direction in most areas is between WNW and NW with sometimes a rather high steadiness (0.3 - 0.5). In the period from September to December in most areas the prevailing wind direction is between SW and W. It is remarkable that in the decade 1971-1980 the steadiness in October is much less and in November and December much greater than in the decade 1961-1970. In January the prevailing wind direction is mostly SSW. February and March are transition months with prevailing winds from several directions. In April and May northerly directions are prevailing, especially in the period 1971-1980. In November, December and January the median wind speed (the speed exceeded in 50% of the observations) mostly falls in the speed interval of force 5 of the Beaufort Scale, in June mostly in force 3 and in the rest of the months in force 4. The differences between northern and southern areas are small.

3.8.2 Exceedance percentages
For each one degree square of the North Sea between 51°N and 60°N monthly percentages of wind forces of respectively greater than Beaufort 5, 6, 7, 8 and 9 and less than Beaufort 3, 4 and 5 have been determined for the period 1961-1980. In figures 33-44 a selection is given in the form of charts with frequencies of wind forces of greater than 5 and 7 Beaufort and less than 4 Beaufort for January, April, July and October. These months appear to be representative for the four seasons. The upper figure in each one degree square gives the frequency and the lower figure the number of observations on which the calculation is based. Some isolines of equal frequencies have been drawn. As there are sometimes large differences between the numbers of observations available for each one-degree square the data have to be considered with some care. The influence of land where frictional effects on the wind are greater than of the sea, is reflected in the pictures. In the charts with wind forces greater than 5 or 7 Beaufort the maximum

frequencies occur in the centre of the northern part of the area while the frequencies gradually decrease towards the coasts. For example the January chart for wind force >7 (gale force or more), figure 34, shows a maximum greater than 20%, while the percentages decrease to about 10% near the British and Norwegian coasts. They also decrease to less than 5% in southern directions but with a secondary maximum of greater than 5% in the southern North Sea. Probably this is caused by the tongue of warmer water coming through the English Channel into the North Sea. Also the prevailing SW-ly wind, of which the fetch is relatively long, can contribute to this phenomenon. The January chart for windforce < 4 Beaufort given in figure 35 shows the opposite with a minimum in the centre of the northern part of the area and an increase of the percentages towards the coasts and to the south. Now a secondary minimum occurs over the warmer water in the southern North Sea.

During the entire year the maximum frequency of gale force occurrences (> 7 Beaufort) is situated in the centre of the northern part of the area. In January the highest isoline is 20%; in February only 10%, in March 15%, April 5%, May 2%, while June is the quietest month with only 1%, in July and August 2%, September 5%, October 10%, November 15% and December 20%. The secondary maximum in the southern North Sea is only present in January and February.

The monthly and annual frequencies of gales (windforce 8 or more), strong winds (force 6 or more) and light winds (force 2 or less) for 8 selected areas and 4 light-vessels (indicated in figure 1) are given in figure 45. Gales are most frequent in November, December and January and least frequent from May to August. During the summer months the differences in occurrence between the northern and southern areas are small. In the winter months, however, the frequency in the north with percentages of 10-20% is much greater than in the south with percentages of about 5-10%. Remarkable is that in the areas north of 56°N December is the windiest month, while south of 56°N this is November. In most areas June is the quietest month.

3.8.3 Wind roses

Information about wind force depending of the wind direction is given in figures 46-57. For three selected areas (areas 03, 15 and 24 in figure 1) wind roses are shown for January, April, July and October over the period 1961-1980. Roses are given for 4 classes of wind force: Beaufort 1-3, Beaufort 4-5, Beaufort 6-7 and Beaufort 8-12. The percentage of calms is given in the centre of the first rose. The roses clearly show that all directions occur. At higher wind force however, there are clearer preferences for certain directions than in the case of light winds. In the southern areas directions between SW and NW are occurring the most (SW in the winter months, while in the summer months NW is occurring more frequently). The frequency of N-ly and NE-ly winds is relatively high in April, May and June and low in October, November and December. SE-ly wind directions occur least, especially from April to August. This picture changes gradually with increasing latitude. SE-ly directions are getting higher percentages, especially during the winter months, while from April to August NW-ly and N-ly directions have relatively high percentages. During October, November and December directions between SW and NW score high. In the northern area NE-ly wind directions occur least.

3.8.4 Persistence diagrams
For the Netherlands lightvessels the duration (persistence) of periods with certain wind conditions ($\geq 6, 7, 8, 9$ and 10 Beaufort and $\leq 2, 3, 4,$ and 5 Beaufort) have been determined. As examples in figures 58-61 persistence diagrams are given for lightvessel Texel (position 53°01'N, 4°22'E) for both the winter season (December, January and February) and the summer season (June, July and August). Examples illustrating the use of the diagrams have been given with the diagrams. From the diagrams several conclusions may be drawn, for instance that on average during the winter season about once a year a period occurs with wind force 8 or more lasting at least a day. Also in the summer season such a situation sometimes occurs, but then only about once in 10 years. The diagrams do not show that it are not the winter months December, January and February that are the windiest, but actually November, December and January. In February the number of periods with wind force 8 or more is almost half that of January. The longest uninterrupted period with wind force 9 or more has been more then 1 day, but less than 1.5 day at Goeree and Terschellingerbank and more than 1.5 days but less than 2 days at Noord Hinder and Texel. The decrease of the numbers with the duration is fast, practically exponential for the shorter durations for a period of 24 hours and after this period somewhat slower. Looking at light winds we see for example that in an average summer season only six times a year a period occurs with wind force 2 or less, persisting a day or longer. In an average winter season such a situation occurs only two times a year.

3.8.5 Probabilities of extreme wind speeds
For planning and design purposes the return period is often used instead of the probability of exceedance. Generally, the probability that the wind speed value U_N with a return period of N years is exceeded equals $1/N$. In many applications $N = 50$ or $N = 100$.

The thus defined return period must be applied with some caution as it is the average value of the length of the time interval between two successive exceedances of U_N. When the occurrence times of the exceedances of U_N are randomly distributed, then the probability that U_N is exceeded in a period of N years equals 63%; the probability of two or more exceedances equals 26%.

In the table below for the Netherlands lightvessels the values with 10, 50 and 100 year return periods are given. The values can be interpreted as hourly mean windspeeds.

return period in years	Goeree m/s	Goeree kts	Noord Hinder m/s	Noord Hinder kts	Texel m/s	Texel kts	Terschellingerbank m/s	Terschellingerbank kts
			wind speed (20 m above sea surface)					
10	29	56	29.5	57	30	58	30	58
50	32	62	32.5	63	33.5	65	33.5	65
100	33	64	33.5	65	34.5	67	34.5	67

The data have been determined from a cumulative frequency distribution of all three-hourly wind speeds during the observation periods considered, irrespective of direction or season. These periods are for Goeree: 1949-1970; Noord Hinder: 1953-1980; Texel: 1949-1977; Terschellingerbank: 1949-1975. The equivalent values (which correspond to an anemometer height of 20 metres above sea surface) belonging to the so- called scientific scale have been assigned to the estimates according to the Beaufort scale (see appendix 2). After this the return values have been estimated by extrapolation on Weibull diagram paper. In the same way also extreme hourly mean windspeeds in m/s with return periods of respectively 10,50 and 100 years have been determined for each one-degree-square of the North Sea between 51°N and 60°N. The extremes have been determined from the cumulative frequency distributions of all available wind speed observations made by selected ships over the period 1961-1980, irrespective of direction or season. The results are given in figures 62, 63 and 64. As could be expected the highest values occur in the centre of the northern part of the area. The values decrease towards the coasts and the south. In the figures no indication of possible errors in the values is given. The spread in the values ranges from about ± 3-4% for the return period of 10 years and to about ± 4-5% for return periods of 50 and 100 years.

The data for the lightvessels are based on observations every three hours. Because the sequential data are not statistically independent, it is not allowed to assume that the duration of an exceedance is equal to the sampling period. This assumption would lead to an over-estimate of the return value U_N. To avoid this there is chosen for a standard exceedance time for extreme values of 4.5 hours. After extrapolation of known durations of less extreme conditions this seems a realistic estimate.

3.8.6 A comparison of the wind data of the Netherlands lightvessels with other periods

The periods which the successive treatises of the observations of the Netherlands lightvessels refer to, have been determined by non-climatological factors. The first treatise of Dr. J.P. van der Stok (1912) bears upon the beginning period of the observations, in which he still made distinction between two sub-periods. The first one was 1859-1883, in which period observations were made at Noord Hinder only. Hereafter observations were made at Schouwenbank, Haaks and Terschellingerbank as well. Van der Stok ended his analysis with the year 1909. For this reason G. Verploegh (1956-1959) chose the year 1910 as a starting point for his study. In September 1939 the lightvessels were withdrawn when the Second World War broke out, ending the period treated by Verploegh, which had also some interruptions in the First World War (1914-1918). Some years after the Second World War the lightvessels had been laid out again, some with different names and at different positions. Because from 1949 onwards visual estimates of wave direction, -period and -height have been incorporated in the observations that year has been chosen as the starting point of the last treated period presented by C.G. Korevaar (1987). The last year 1980 has been chosen, because the traditional lightvessel-observations had come to an end in this year. In this way four periods have been created, all with a length of 25-30 years in which the observations have been made in about the same way. In the table of figure 65 frequencies of wind force ≥ 4, ≥ 6, ≥ 8 and ≥ 10 Beaufort have been given for the periods mentioned and for the different lightvessels, regardless of direction. Only the lightvessels Noord Hinder and Terschellingerbank have occupied approximately the same position during the whole period. Because their positions are close enough, the data of Goeree may be compared with those of the former lightvessel Schouwenbank and of Texel with those of Haaks.

If we look at wind force 4 and more we see that obvious differences are existing between the periods considered. At lightvessel Noord Hinder for example the annual frequency was 61% in the first period (1859-1883). In the second period (1884-1908) this percentage was 50% only and in the third period (1910-1939) even 45%, while in the last period (1953-1980) the percentage was 59%, about as much as in the first period.

For the other lightvessel positions the first period is not available, but the other three periods give the same picture with lower frequencies in the second and the third period and a higher frequency in the fourth. At higher wind forces this picture changes. The frequencies come closer together for all periods except for the first one. Looking at the post war period only, it strikes that the mean annual frequency of wind force 4 or more is the same at all four positions. The differences for the greater wind forces are also small in the last period. Texel shows the highest annual percentage (2.7%) with wind force 8 or more. The annual frequency (0.2%) of windforce 10 or more is the same again on all positions.

For lightvessels Haaks (1891-1938) and Texel (1949-1980), which occupied about the same positions the numbers of storms (wind force 10 or more) in each year are given in figure 66 (due to the Second World War no data are available for the period 1939-1947). For some years the data have been substituted by those of other lightvessels, for example during the First World War by those of lightvessels Maas and after 1977 by those of lightvessel Noord Hinder.

From figure 66 it appears that there have been years, although only a few, without the occurrence of a storm (windforce ≥ 10). The highest number of storms was 7 (1908, 1909 and 1949). During the whole period of 81 years 217 storms (on an average 2.7 a year), have raged about with an average duration of 6-8 hours. The longest uninterrupted period with wind force 10 or more had a duration of 30 hours, which occurred in February 1962 and January 1976.

The number of storms is not spread regularly over the years. It is striking that there are groups of years with obviously more storms than average, while there are also periods in which only few storms have occurred. It is not possible to discover a clear periodicity in this. When considering longer periods it can be said that during the period 1900-1930 about twice as many storms occurred with regard to the period 1951-1980. There are obvious differences in the various seasons. From the 217 storms 112 occurred in winter (47 in December, 42 in January and 23 in February), 84 in autumn (8 in September, 32 in October and 44 in November), 13 in spring (10 in March, 2 in April and 1 in May) and 8 in summer (0 in June, 4 in July and 4 in August).

As to the wind directions during storms 75% were westerly (SW to NW), 14% northerly (NNW, N, NNE), and 8% southerly (S, SSW). The directions E and ESE count together as 3%. From NE and ENE came only 0.4%, while no storm occurred with a wind direction of SSE or SE.

3.9 Waves

3.9.1 General remarks
Except for swell the directional distribution of the waves corresponds to that of the wind. Differences occur in the roses. This is caused by the fact that in making the wave roses from each observation the highest system (either sea or swell) has been used. In the case of equal height the system with the longer period was chosen. The height of the sea waves corresponds to the wind speed. Limiting factors for the growth of the waves are duration, fetch and water depth. In the North Sea only NW-ly winds may have very long fetches. If the duration in this situation is long enough very high waves develop, but the height of the waves is more often limited by the short duration of the wind from this direction. The limiting factor with winds from other directions is mostly the fetch. Especially in the southern North Sea the water depth is another limiting factor. The annual median wave heights (the median wave height is the height exceeded in 50% of the observations) vary between 1 and 2 m in the north and between 0.5 and 1 m in the south. Especially in the north the values are slightly higher during the period 1971-1980 than during the period 1961-1970.

3.9.2 Exceedance percentages
For each one-degree-square (a square of one degree latitude by one degree longitude) of the North Sea between 51°N and 60°N monthly percentages of wave heights of respectively greater than 1.75, 2.75, 3.75, 4.75 and 5.75 metres and less than 0.25, 0.75 and 1.25 metres have been determined for the period 1961-1980. From each observation the highest of sea or swell has been used. In figures 67-78 a selection is given in the form of charts with frequencies of wave heights greater

than 1.75 and 3.75 m and less than 1.25 m for January, April, July and October, which months can be thought to be representative for the 4 seasons. The upper figure in each one-degree-square gives the frequency and the lower figure the number of observations on which the calculation is based. Some isolines of equal frequencies have been drawn. As there are sometimes large differences between the numbers of observations available for each one-degree-square the data have to be interpreted with some care. Both because the wind is stronger in the open sea and because of the longer fetches in the charts with wave heights greater than 1.75 m and 3.75 m the maximum frequencies occur in the centre of the northern part of the area while the frequencies gradually decrease towards the coasts. For example the January chart for wave heights of greater than 3.75 m, given in figure 68 shows a maximum of more than 30%, while the percentages decrease to about 10% near the British and Norwegian coasts. They also decrease southwards to less than 5%, but with a secondary maximum south of the Dogger Bank of more than 5%. The January chart for wave heights of less than 1.25 m shows the opposite with a minimum in the centre of the northern part of the area and an increasing of the percentages to the coasts and to the south, while a secondary maximum occurs at the Dogger Bank. During the whole year the maximum frequency of wave heights of greater than 3.75 m occurs in the centre of the northern part of the area. In January the highest isoline is 30%, in February only 20%, in March 30%, April 10%, May 5%, June 5%, July 5%, August 5%, September 10%, October 20%, November 25% and December 30%. The monthly and annual frequencies of wave heights ≥ 4 m and ≥ 6 m for 8 selected areas and 4 lightvessels (indicated in figure 1) are given in figure 79. There is a well-marked annual variation with the highest frequencies from October to March and the lowest frequencies from May to August. It is remarkable that in many areas a secondary maximum (and sometimes the real maximum) occurs in March. There are large differences between the north and the south with of course the greater percentages in the north. In the same figures the frequencies of low waves (heights ≤ 1.5m) are given.

From the visually estimated wave parameters the period is the least reliable. Therefore the only period data given (also in figure 79) are the monthly and annual percentages of observations with periods ≥ 6 seconds. The choice of ≥ 6 seconds is more or less arbitrary but is related to the fact that in the reporting code all periods ≤ 5 seconds were indicated by the same code value. The annual variation shows a similar pattern as for the wave heights ≥ 4 and ≥ 6 m.

3.9.3 Wave roses
Information about wave height depending on wave direction is given in figures 80-91. For three selected areas (areas 03, 15 and 24 in figure 1) wave roses are shown for January, April, July and October over the period 1961-1980. Roses are given for 4 classes of wave height in 0.5 metre values, namely 0-3 (0-1.5 m), 4-7 (2-3.5 m), 8-11 (4-5.5 m) and greater than 11 (5.5 m). The percentage of observations with a calm sea is given in the centre of the first rose. From each observation the highest system (either sea or swell) has been used. In the case of equal

heights, the system with the longer period was chosen. The higher waves show a clearer preference for certain directions than the lower waves, in analogy with wind directions at greater wind force. In the southern areas, in the case of the higher waves, NW-ly directions are occurring the most frequent, while more to the north also SE-ly directions are occurring more. In contrast with the prevailing wind and sea direction, the prevailing swell is almost always from a northerly direction. This is shown in figure 92, in which the prevailing monthly and annually swell directions are given for 8 selected areas (figure 1) during the two decades 1961-1970 and 1971-1980. In figure 92 the prevailing swell direction is given by the resultant vector mean direction with each individual vector length set equal to 1.

3.9.4 Persistence diagrams
For the Netherlands lightvessels the duration (persistence) of periods with certain wave conditions (> 0.75, 1.75, 2.75, 3.75 and 4.75m, and < 0.25, 0.75, 1.25, 1.75 and 2.25 m) have been determined. As examples in figures 93-96 persistence diagrams are given for lightvessel Texel (position 53°01'N, 4°22'E) for both the winter season (December, January and February) and the summer season (June, July and August). Examples illustrating their use have been given with the diagrams.

The conclusions which may be drawn from the diagrams are similar to those of the wind conditions (cf. par. 3.8.4). For instance in an average winter season about 2-3 times a year a period occurs with a wave height greater than 1.75 m and a duration of at least two days. Also in the summer season such a situation sometimes occurs, but much less, namely at Noord Hinder about once in 5 years, at Texel and Terschellingerbank about 2 times in 5 years and at Goeree about 4 times in 5 years.

During an average summer season about 7-9 times a year a period occurs with a wave height of less than 0.75 m and a duration of at least two days. In winter this situation occurs twice as less as in summer. It has occurred both in winter and in summer that this wave condition lasted more than two weeks.

3.9.5 Probabilities of extreme wave heights
Of course measurements are more suitable than estimates for determining extreme wave heights. However, sufficiently long series of measurements are hardly available. In general it is the opinion that the statistics based on large enough numbers of visual observations are sufficiently reliable. A matter of concern related with estimating extreme wave heights is the relation between significant wave height H_s from measured records and the visually estimated wave height H_v. The relation between H_v and H_s has been studied by several authors. In case of low waves values for the ratio H_s/H_v are usually given between 1.0 and 1.1. For higher waves the following relationships have been found, based on weathership data:

$H_s = 1.68\ H_v^{0.75}$ (Nordenstrøm, 1969)

or $H_v = 0.98\ H_s + 0.5$ (Jardine 1979)

or $H_s = 2.33 + 0.75\ H_v$ (Soares, 1986).

Such relations must be looked at with some caution. For example for $H_v = 10$ m according to Nordenstrøm, $H_s = 9.4$ m, according to Jardine $H_s = 9.7$ m and according to Soares $H_s = 9.8$ m. These results are close together, but for $H_v = 15$ m there are already great differences, with respectively 12.8 m, 14.8 m and 13.6 m. This suggests a certain overestimation in extreme conditions. On the other hand, such relations only refer to observations, not to extrapolations like the return period values as given in figures 97-99. Another source of deviations is the difference of the height of the observer's eye above the water surface.
On a weathership this is usually about 6 m or more, while at a lightship this may be 4 m or less. Bouws (private communication) suggests an alternative relation for lightship observations:

$H_s = 1.4\ H_v^{0.75}$

implying an overestimation of extreme wave heights of about 1 metre. Another complicating factor is the increasing influence of bottom depth on extreme wave heights, suggesting that straightforward extrapolation of the observed distribution of wave heights does not hold fully for shallow areas like the southern North Sea (see Bouws, 1978).
In the table below the values with 10, 50 and 100 years return periods of extreme wave heights are given based on visually estimated wave heights (H_v) in metres.

return period in years	wave height in metres			
	Goeree	Noord Hinder	Texel	Terschellingerbank
10	6.5	6.6	6.5	6.5
50	7.3	7.5	7.3	7.4
100	7.7	7.8	7.7	7.8

The data have been determined from a cumulative frequency distribution of all three-hour wave-height-observations during the observation periods considered, irrespective of direction or season. These periods are for Goeree: 1949-1970; Noord Hinder: 1953-1980; Texel: 1949-1977; Terschellingerbank: 1949-1975. The return periods have been estimated by extrapolation on Weibull diagram paper. In compiling the cumulative frequency distributions from each individual observation the highest of sea or swell has been used.
In the same way also extreme wave heights in metres with return periods of

respectively 10, 50 and 100 years have been determined for each one-degree-square of the North Sea between 51°N and 60°N. The data have been determined from the cumulative frequency distributions of all available visually estimated wave heights by voluntary observing ships over the period 1961-1980, irrespective of direction or season. The return periods have been estimated by extrapolation on Weibull diagram paper. In making the cumulative frequency distributions from each individual observation the highest of sea or swell has been used. The results are given in figures 97,98 and 99. As could be expected the highest values occur in the centre of the northern part of the area. The values decrease towards the coasts and the south. In the figures no indication of possible errors is given. For the return period of 10 years the spread in the values ranges from about ± 7% in the south to about ± 12% in the north. For the return periods of 50 and 100 years the spread is about ± 8% in the south and ± 13-14% in the north.

The data for the lightvessels are based on observations every three hours. Because the sequential data are not statistically independent, it is not allowed to assume that the duration of an exceedance is equal to the sampling period. This assumption would lead to an over-estimate of the return value. To avoid this there is chosen for a standard exceedance time for extreme values of 4.5 hours. After extrapolation of known durations of less extreme conditions this seems a realistic estimate.

References

E. *Bouws (1978)*. Wind and wave climate in the Netherlands sector of the North Sea between 53° and 54° north latitude. KNMI scientific reports WR 78-9.

R. *Dorrestein (1967)*. Wind and wave data of Netherlands lightvessels since 1949. KNMI Mededelingen en Verhandelingen 90. 's-Gravenhage.

J.M. *Dury (1970)*. The Beaufort scale of wind force. WMO Reports on Marine Science Affairs No. 3, Genève.

A.E. *Graham (1982)*. Winds estimated by the Voluntary Observing Fleet compared with instrumental measurements at fixed positions. The Meteorological Magazine 111: 312-327.

T.P. *Jardine (1979)*. The reliability of visually observed wave heights. Coastal engineering 3: 33-38.

L. *Kaufeld (1981)*. The development of a new Beaufort equivalent scale. Meteorologische Rundschau 34: 17-23.

C.G. *Korevaar (1987)*. Climatological data of the Netherlands lightvessels over the period 1949-1980. KNMI scientific reports WR 87-9.

C.G. *Korevaar (1989)*. Climatological data for the North Sea based on observations by voluntary observing ships over the period 1961-1980. KNMI scientific reports WR 89-02.

A.K. *Laing (1985)*. An Assessment of Wave Observations from Ships in Southern Oceans, Journal of Climate and Applied Meteorology, 24: 481-494.

N. *Nordenstrøm (1969)*. Methods for predicting long-term distributions of wave loads and probability of failure for ships, Appendix II: Relationships between visually estimated and theoretical wave heights and periods. Rep. 69-22 S, Det Norske Veritas.

C.G. *Soares (1986)*. Assessment of the uncertainty in visual observations of wave height, Ocean Engineering, 13: 37-56.

J.P. *v.d. Stok (1912)*. Das Klima des südöstlichen Teiles der Nordsee, unweit der Niederländischen Küste. KNMI Mededelingenen verhandelingen 13 a, b, c, Utrecht.

F.S. *Terziev (1981)*. Investigation of contemporary methods of measuring sea surface and surface-layer temperatures. WMO Marine Meteorology and related oceanographic activities Report No. 2.

G. *Verploegh (1956-1959)*. Climatological data of the Netherlands lightvessels over the period 1910-1940. KNMI Mededelingen en verhandelingen 67, Part I: Statistics of gales; Part II: Air pressure and wind; Part III: Temperatures and hydrometeors, thunderstorms, general discussion. 's-Gravenhage.

G. *Verploegh (1956)*. The equivalent velocities for the Beaufort estimates of the wind force at sea. KNMI Mededelingen en Verhandelingen 66, 's-Gravenhage.

G. *Verploegh (1961)*. On the accuracy and the interpretation of wave observations from selected ships. Working paper, CMM Working group for Technical problems. Genève.

G. *Verploegh (1967)*. Observation and analysis of the surface wind over the ocean. KNMI Mededelingen en Verhandelingen 89, 's-Gravenhage.

Appendix 1

Beaufort scale of wind force for reporting wind at sea.

Beaufort number	Descriptive term	Windspeed equivalents metres/sec	knots	Specifications
0	Calm	0- 0.2	<1	Sea like a mirror
1	Light air	0.3- 1.5	1- 3	Ripples with the appearance of scales are formed, but without foam crests.
2	Light breeze	1.6- 3.3	4 - 6	Small wavelets, still short but more pronounced; crests have a glassy appearance and do not break.
3	Gentle breeze	3.4- 5.4	7 -10	Large wavelets; crests beging to break; foam of glassy appearance; perhaps scattered white horse.
4	Moderate breeze	5.5- 7.9	11-16	Small waves becoming longer; fairly frequent white horses.
5	Fresh breeze	8.0-10.7	17-21	Moderate waves, taking a more pronounced long form; many white horses are formed (chance of some spray.)
6	Strong breeze	10.8-13.8	22-27	Large waves begin to form; the white foam crests are more extensive everywhere (probably some spray).
7	Near gale	13.9-17.1	28-33	Sea heaps up and white foam from breaking waves begins to be blown in streaks along the direction of the wind.
8	Gale	17.2-20.7	34-40	Moderately high waves of greater length; edges of crests begin to break into the spin drift; the foam is blown in wellmarked streaks along the direction of the wind.
9	Strong gale	20.8-24.4	41-47	High waves; dense streaks of foam along the direction of the wind; crests of waves begin to topple, tumble and roll over, spray may effect visibility
10	Storm	24.5-28.4	48-55	Very high waves with long overhanging crests; the resulting foam in great patches, is blown in dense white streaks along the direction of the wind; on the whole the surface of the sea takes a white appearance; the tumbling of the sea becomes heavy and shock-like; visibility affected.
11	Violent storm	28.5-32.6	56-63	Exceptionally high waves (small and medium-sized ships might be for a time lost to view behind the waves); the sea is completely covered with long white patches of foam lying along the direction of the wind; everywhere the edges of the wave crests are blown froth; visibility affected.
12	Hurricane	32.7 and over	64 and over	The air is filled with foam and spray; sea completely white with driving spray; visibility very seriously affected.

Appendix 2.

Beaufort scale of wind for use in scientific projects.

Beaufort number	windspeed equivalents metres/sec	knots
0	0 - 1.3	0 - 2
1	1.4 - 2.7	3 - 5
2	2.8 - 4.5	6 - 8
3	4.6 - 6.6	9 - 12
4	6.7 - 8.9	13 - 16
5	9.0 - 11.3	17 - 21
6	11.4 - 13.8	22 - 26
7	13.9 - 16.4	27 - 31
8	16.6 - 19.2	32 - 37
9	19.3 - 22.4	38 - 43
10	22.5 - 26.0	44 - 50
11	26.1 - 30.0	51 - 57
12	30.1 and above	58 and above

Chapter 4
Figures

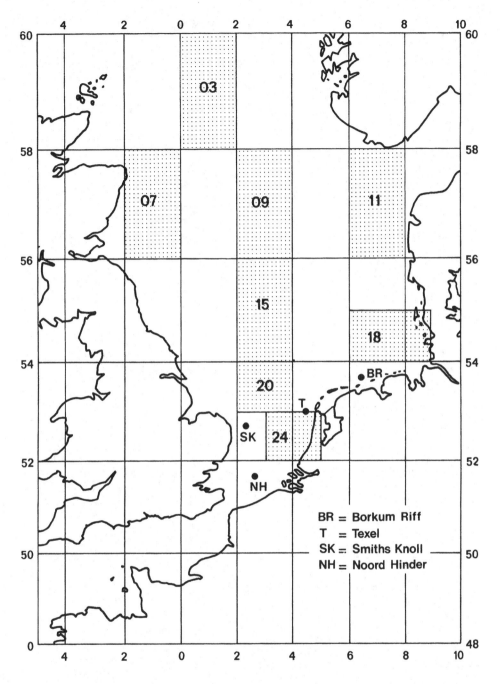

Figure 1 Selected areas and lightvessels for which statistical data are given in figures 10-16, 25-32, 45-61, 65, 66, 79-96.

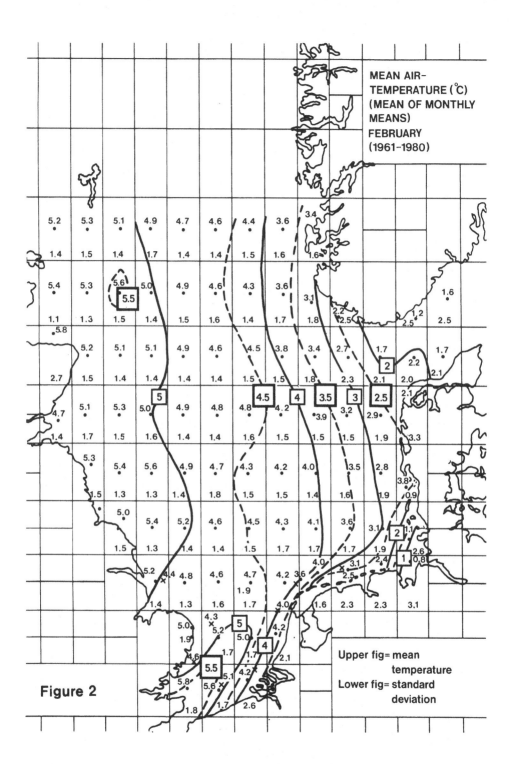

Figure 2

MEAN AIR-TEMPERATURE (°C)
(MEAN OF MONTHLY MEANS)
FEBRUARY
(1961-1980)

Upper fig = mean temperature
Lower fig = standard deviation

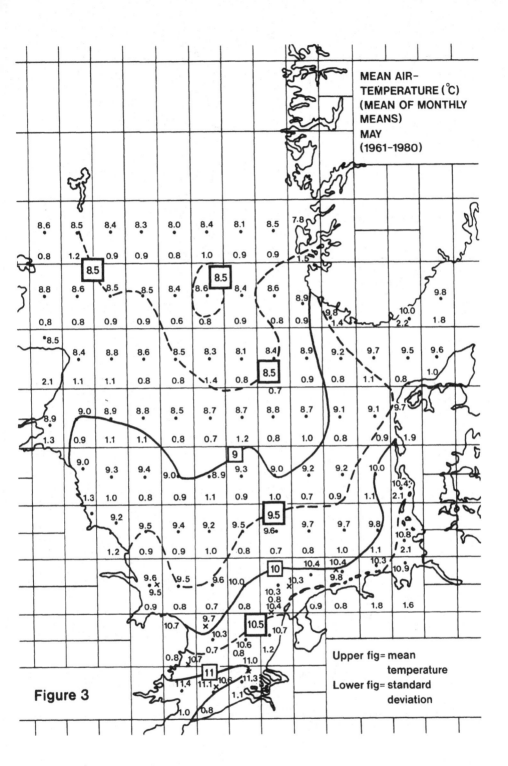

Figure 3

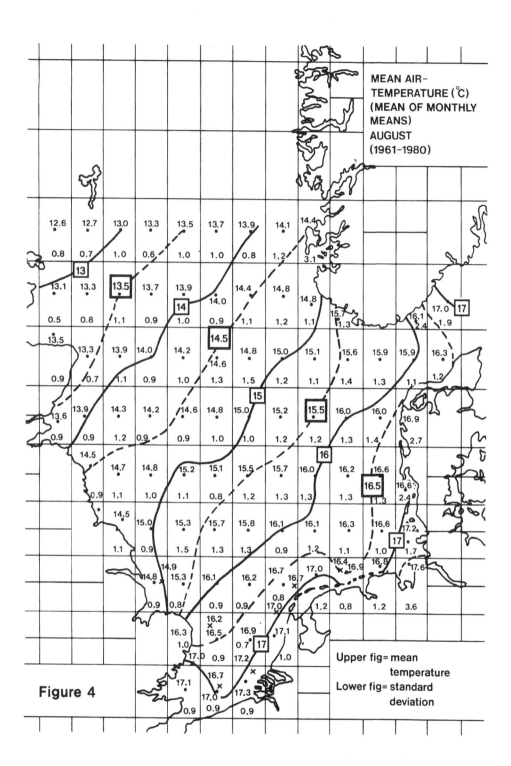

MEAN AIR-
TEMPERATURE (°C)
(MEAN OF MONTHLY
MEANS)
AUGUST
(1961–1980)

Upper fig = mean
temperature
Lower fig = standard
deviation

Figure 4

40

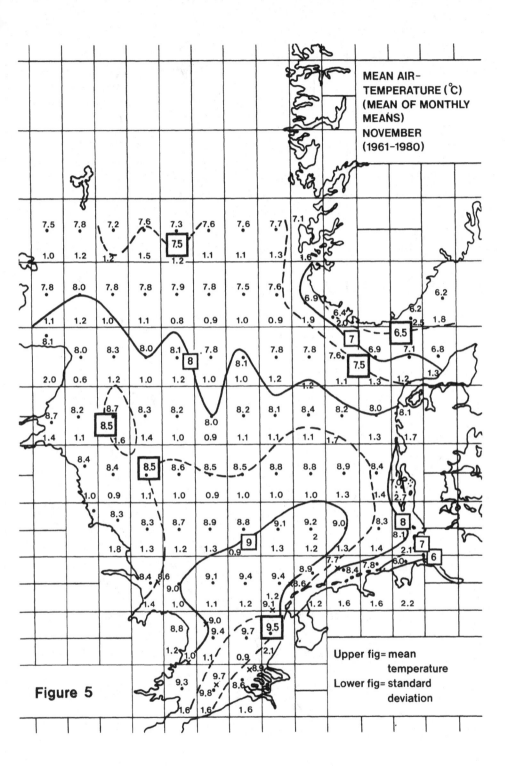

Figure 5

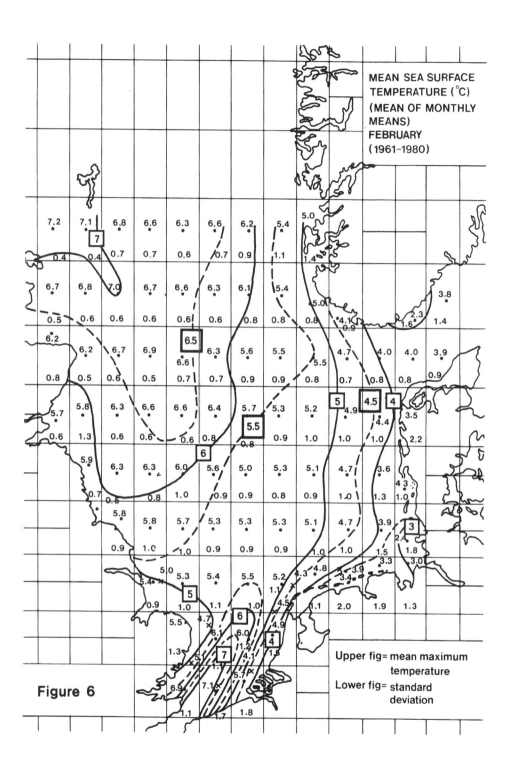

Figure 6

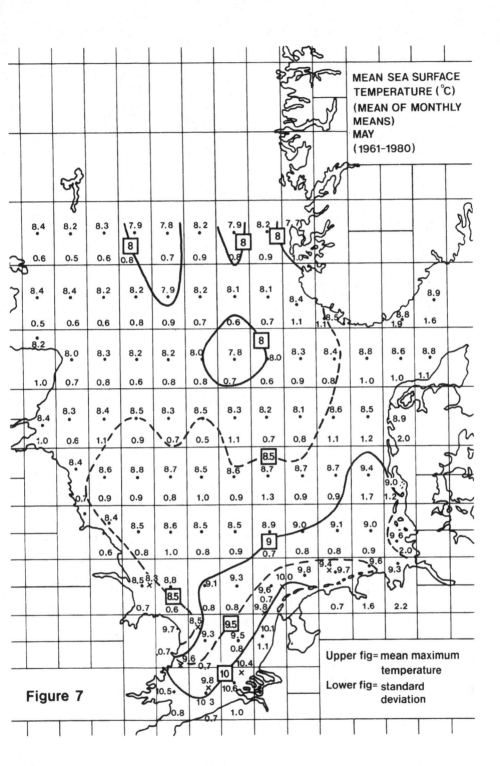

Figure 7

MEAN SEA SURFACE TEMPERATURE (°C)
(MEAN OF MONTHLY MEANS)
MAY
(1961–1980)

Upper fig= mean maximum temperature
Lower fig= standard deviation

43

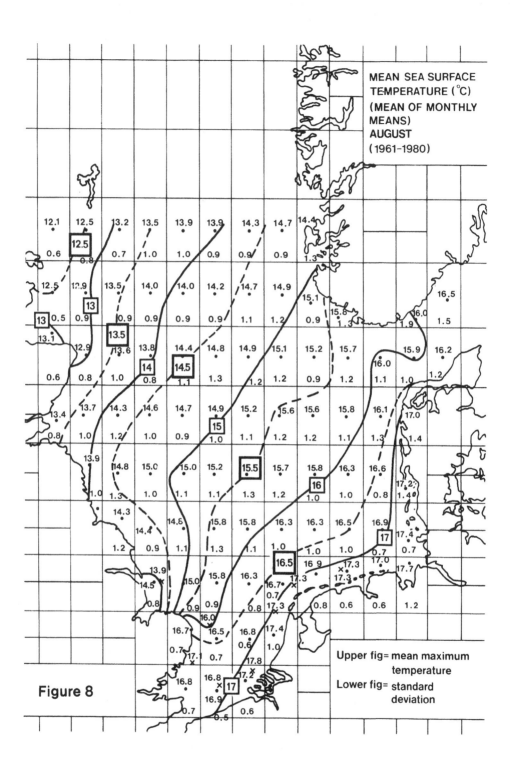

Figure 8

MEAN SEA SURFACE
TEMPERATURE (°C)
(MEAN OF MONTHLY
MEANS)
AUGUST
(1961-1980)

Upper fig= mean maximum
temperature
Lower fig= standard
deviation

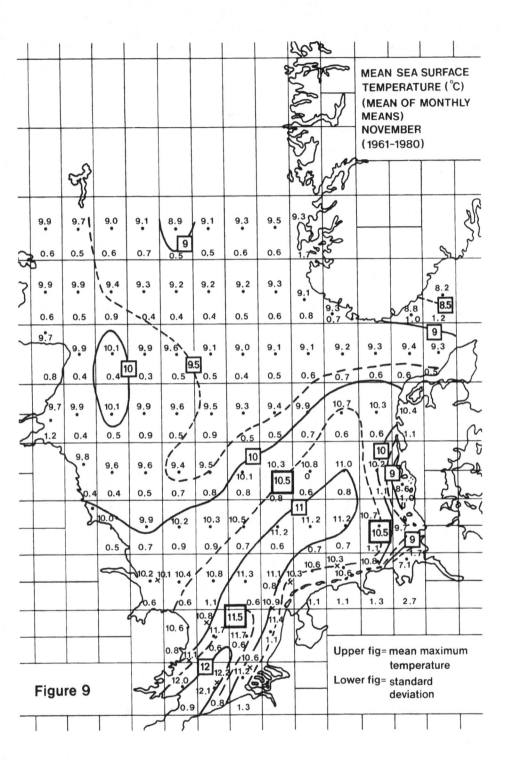

Figure 9

MEAN SEA SURFACE TEMPERATURE (°C)
(MEAN OF MONTHLY MEANS)
NOVEMBER
(1961–1980)

Upper fig= mean maximum temperature
Lower fig= standard deviation

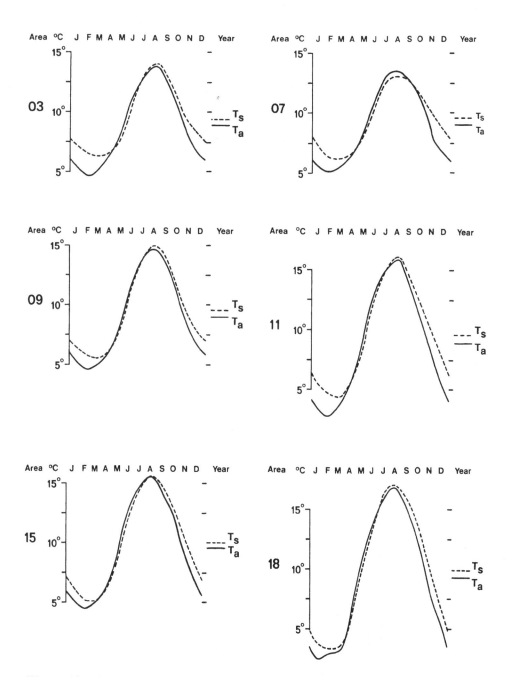

Figure 10a Annual variation of the air temperature (Ta) and sea surface temperature (Ts) for the subareas 03,07,09,11,15 and 18 indicated in figure 1.

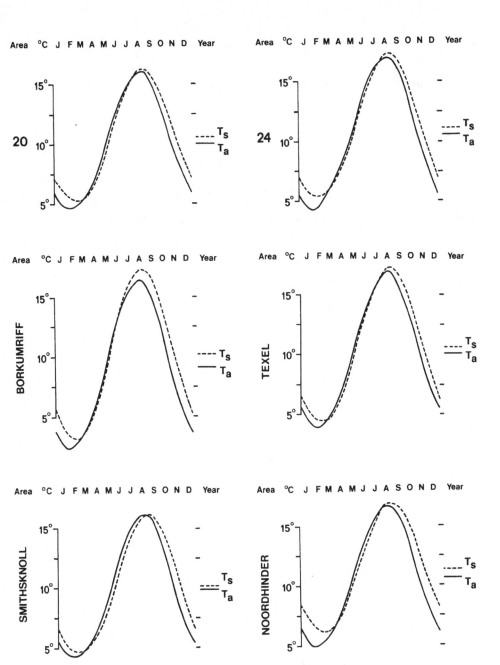

Figure 10b Annual variation of the air temperature (Ta) and sea surface temperature (Ts) for the subareas 20 and 24 and 4 lightvessels indicated in figure 1.

Texel air temperature 1949 - 1977

Numbers of observations for each temperature interval of 1 degree celcius

Temp °C	JAN	FEB	MAR	APR	MAY	JUN	JUL	AUG	SEP	OCT	NOV	DEC	YEAR
≥24.0	-	-	-	-	-	-	1	2	2	-	-	-	5
23.0-23.9	-	-	-	-	-	-	1	14	2	-	-	-	17
22.0-22.9	-	-	-	-	-	4	9	32	3	-	-	-	48
21.0-21.9	-	-	-	-	-	-	31	86	10	-	-	-	127
20.0-20.9	-	-	-	-	-	17	60	173	46	1	-	-	297
19.0-19.9	-	-	-	-	1	23	150	424	140	5	-	-	743
18.0-18.9	-	-	-	-	2	74	423	931	457	14	-	-	1901
17.0-17.9	-	-	-	-	9	209	1024	1659	920	104	-	-	3925
16.0-16.9	-	-	-	1	24	385	1503	1631	1361	338	-	-	5243
15.0-15.9	-	-	-	-	57	759	1821	1277	1390	676	15	-	5995
14.0-14.9	-	-	-	3	147	1274	1198	465	982	1023	39	-	5131
13.0-13.9	-	-	-	6	375	1641	507	68	539	1221	159	-	4516
12.0-12.9	-	-	-	27	797	1378	145	3	234	1016	457	20	4077
11.0-11.9	-	-	5	96	1337	775	58	2	74	839	796	80	4062
10.0-10.9	10	1	12	329	1610	304	1	-	25	589	1233	366	4480
9.0-9.9	115	21	89	591	1385	60	-	-	9	314	1229	787	4600
8.0-8.9	436	192	435	1052	903	37	-	-	-	156	917	1078	5206
7.0-7.9	1030	555	872	1496	336	7	-	-	-	80	593	1072	6041
6.0-6.9	1358	1051	1219	1422	127	2	-	-	-	60	396	990	6625
5.0-5.9	1147	1232	1256	1034	17	-	-	-	-	4	288	705	5683
4.0-4.9	786	987	1015	450	2	-	-	-	-	-	197	503	3940
3.0-3.9	627	654	859	206	-	-	-	-	-	-	119	429	2894
2.0-2.9	463	466	619	55	-	-	-	-	-	-	63	283	1949
1.0-1.9	346	383	401	5	-	-	-	-	-	-	37	200	1372
0.0-0.9	226	275	240	2	-	-	-	-	-	-	21	152	916
-1.0--0.1	175	245	107	-	-	-	-	-	-	-	3	97	627
-2.0--1.1	159	190	46	-	-	-	-	-	-	-	2	75	472
-3.0--2.1	92	125	13	-	-	-	-	-	-	-	-	38	268
-4.0--3.1	52	69	1	-	-	-	-	-	-	-	-	14	136
-5.0--4.1	53	52	-	-	-	-	-	-	-	-	-	2	107
-6.0--5.1	35	15	-	-	-	-	-	-	-	-	-	-	50
-7.0--6.1	14	17	-	-	-	-	-	-	-	-	-	-	31
-8.0--7.1	1	8	-	-	-	-	-	-	-	-	-	-	9
-9.0--8.1	-	-	-	-	-	-	-	-	-	-	-	-	-
≤-9.1	-	-	-	-	-	-	-	-	-	-	-	-	-
TOTAL	7125	6538	7189	6775	7129	6949	6932	6767	6194	6440	6564	6891	81493
MEAN	4.7	4.0	4.9	7.1	10.4	13.5	15.8	17.0	15.7	12.8	9.1	6.4	10.0
ST.DEV.	2.9	2.9	2.3	1.8	1.8	1.8	1.6	1.6	1.8	2.2	2.5	2.9	5.1

Figure 11. Frequency distribution of the air temperature with mean and standard deviation per month and for all months together for lightvessel Texel.

Numbers of observations for each temperature interval of 1 degree celcius

Temp °C	JAN	FEB	MAR	APR	MAY	JUN	JUL	AUG	SEP	OCT	NOV	DEC	YEAR
≥22.0	-	-	-	-	-	-	-	-	-	-	-	-	-
21.0-21.9	-	-	-	-	-	-	1	5	-	-	-	-	6
20.0-20.9	-	-	-	-	-	-	5	23	-	-	-	-	28
19.0-19.9	-	-	-	-	-	2	29	251	57	-	-	-	339
18.0-18.9	-	-	-	-	-	5	223	1178	429	-	-	-	1835
17.0-17.9	-	-	-	-	-	20	1000	2755	1676	26	-	-	5477
16.0-16.9	-	-	-	-	-	124	1986	1949	2421	515	-	-	6995
15.0-15.9	-	-	-	-	-	644	2141	591	1340	1405	-	-	6121
14.0-14.9	-	-	-	-	-	1367	1209	-	264	1591	26	-	4457
13.0-13.9	-	-	-	-	24	1996	271	-	6	1865	298	-	4460
12.0-12.9	-	-	-	-	309	1696	60	-	-	784	1154	-	4003
11.0-11.9	-	-	-	2	1417	794	-	-	-	177	1605	22	4017
10.0-10.9	-	-	-	19	1640	282	-	-	-	70	1781	270	4062
9.0-9.9	-	-	-	240	1645	20	-	-	-	3	1205	1109	4222
8.0-8.9	88	-	30	971	1370	-	-	-	-	3	368	1912	4742
7.0-7.9	926	20	417	1624	571	-	-	-	-	-	101	1840	5499
6.0-6.9	2149	892	1120	1915	146	-	-	-	-	-	30	1085	7337
5.0-5.9	2082	1845	1915	1101	7	-	-	-	-	-	-	512	7462
4.0-4.9	1388	2278	1708	721	-	-	-	-	-	-	-	133	6228
3.0-3.9	300	800	1262	172	-	-	-	-	-	-	-	13	2547
2.0-2.9	103	232	514	9	-	-	-	-	-	-	-	-	858
1.0-1.9	44	195	110	-	-	-	-	-	-	-	-	-	349
0.0-0.9	70	175	104	-	-	-	-	-	-	-	-	-	349
-1.0- -0.1	39	105	9	-	-	-	-	-	-	-	-	-	153
-2.0- -1.1	-	2	-	-	-	-	-	-	-	-	-	-	2
< = -2.1	-	-	-	-	-	-	-	-	-	-	-	-	-
TOTAL	7189	6544	7189	6774	7129	6950	6925	6752	6193	6439	6568	6896	81548
MEAN	5.6	4.5	4.8	6.6	9.8	13.3	15.9	17.2	16.6	14.2	10.8	7.8	10.5
ST.DEV	1.4	1.5	1.5	1.4	1.4	1.4	1.2	0.9	1.0	1.3	1.3	1.3	4.7

Figure 12. Frequency distribution of the sea surface temperature with mean and standard deviation per month and for all months together for light-vessel Texel.

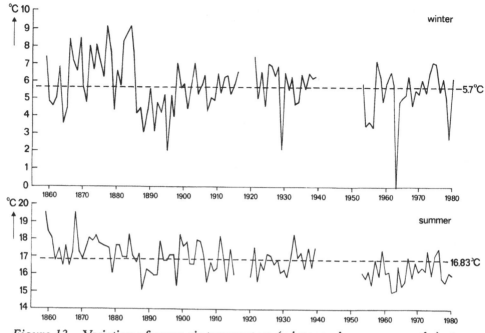

Figure 13 Variation of mean air temperature (winter and summer months) at Noord Hinder (1859-1980).

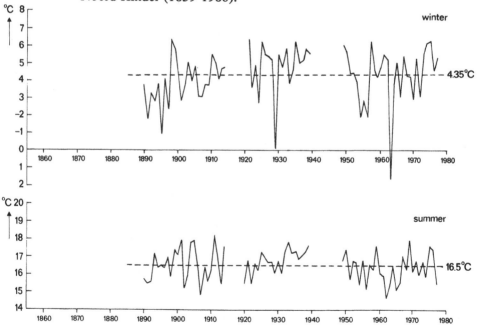

Figure 14. Variation of mean air temperature (winter and summer months) at Haaks/Texel (1890-1977).

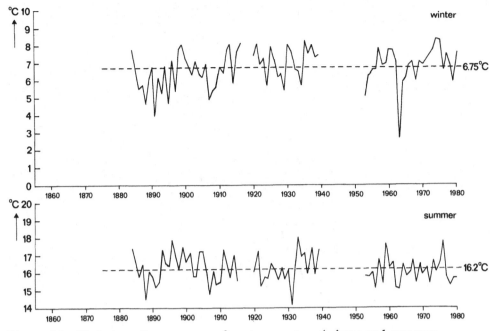

Figure 15   Variation of mean sea surface temperature (winter and summer months) at Noord Hinder (1885-1980).

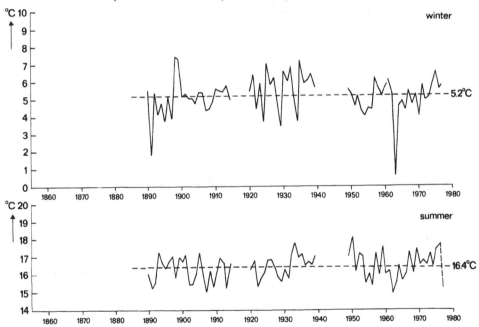

Figure 16   Variation of mean sea surface temperature (winter and summer months) at Haaks/Texel (1890-1977).

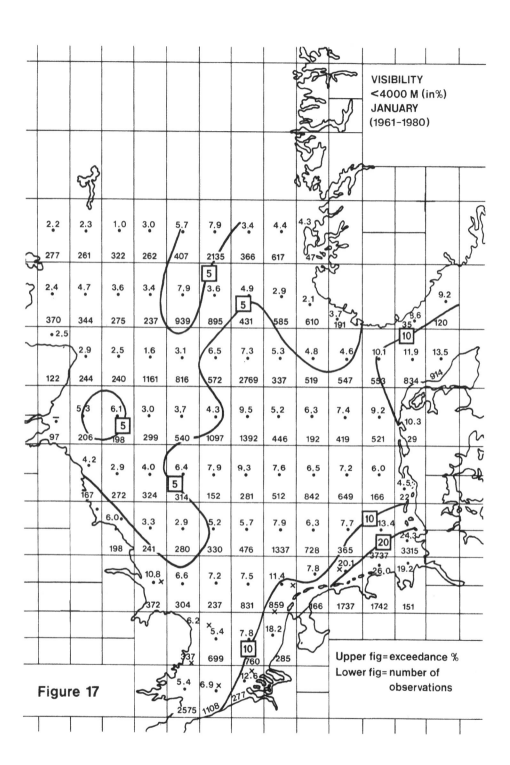

Figure 17

VISIBILITY
<4000 M (in%)
JANUARY
(1961-1980)

Upper fig = exceedance %
Lower fig = number of observations

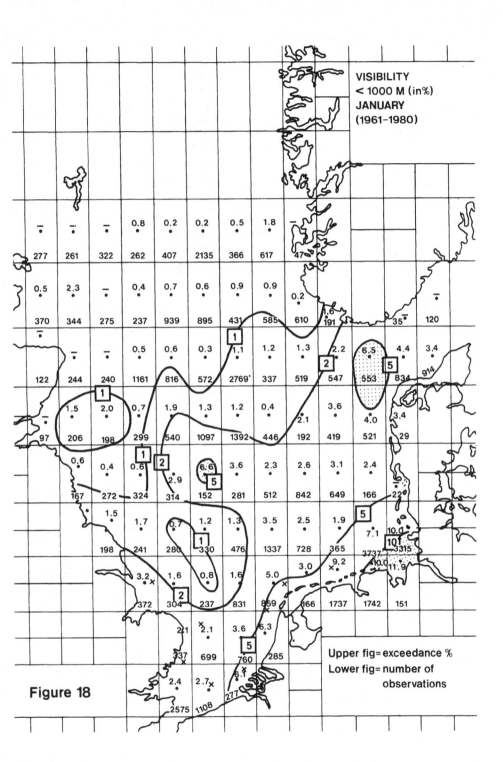

Figure 18

VISIBILITY
< 1000 M (in%)
JANUARY
(1961–1980)

Upper fig= exceedance %
Lower fig= number of
observations

53

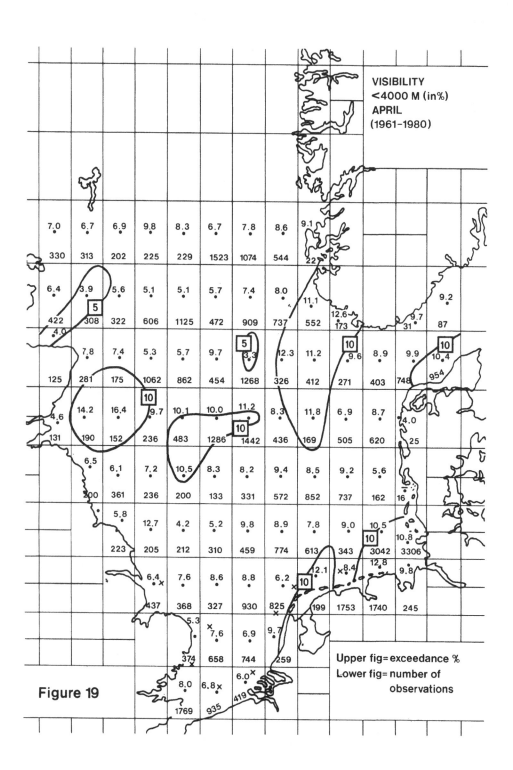

Figure 19

VISIBILITY
<4000 M (in%)
APRIL
(1961–1980)

Upper fig= exceedance %
Lower fig= number of observations

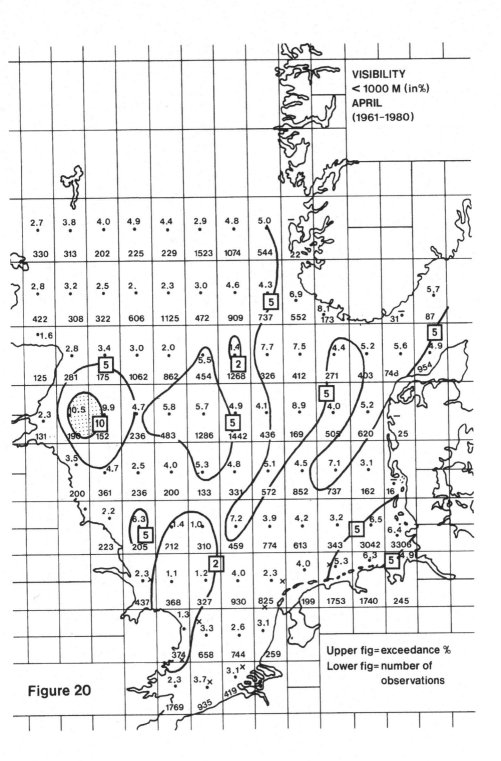

Figure 20

VISIBILITY
< 1000 M (in%)
APRIL
(1961-1980)

Upper fig=exceedance %
Lower fig= number of
observations

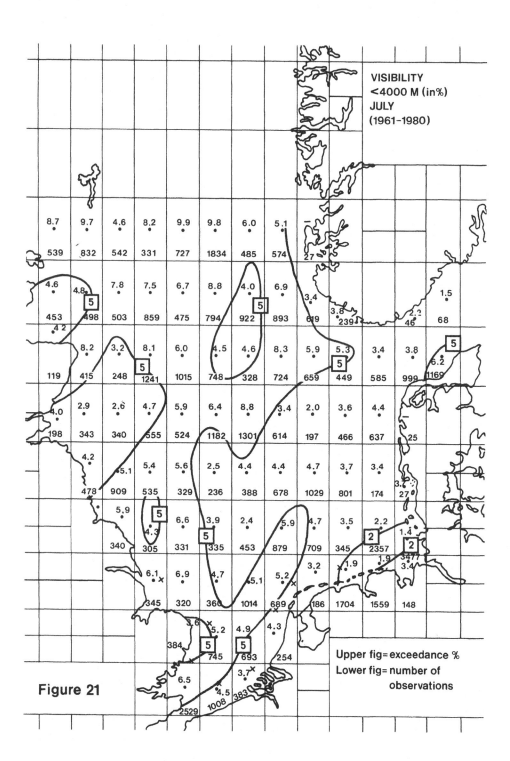

VISIBILITY
<4000 M (in%)
JULY
(1961-1980)

Upper fig = exceedance %
Lower fig = number of
observations

Figure 21

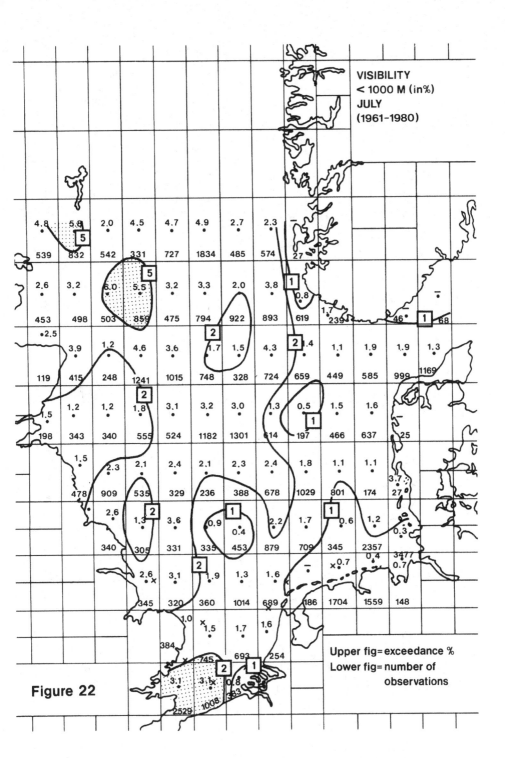

Figure 22

VISIBILITY
< 1000 M (in%)
JULY
(1961-1980)

Upper fig=exceedance %
Lower fig= number of
observations

57

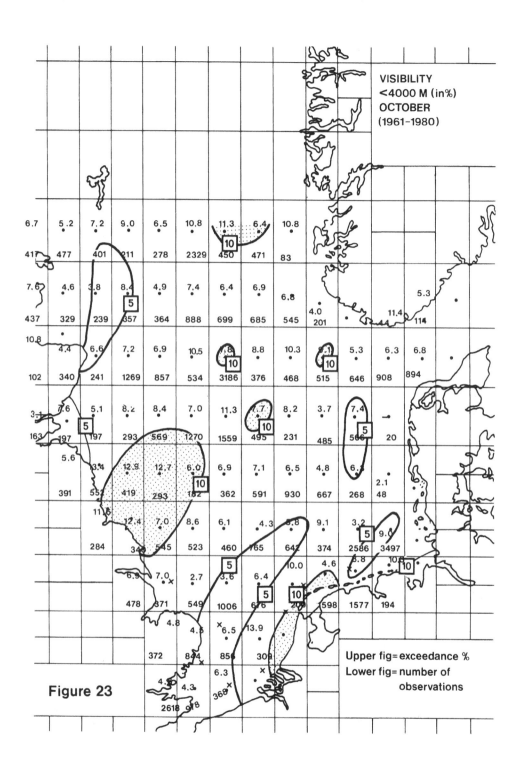

Figure 23

VISIBILITY
<4000 M (in%)
OCTOBER
(1961–1980)

Upper fig = exceedance %
Lower fig = number of observations

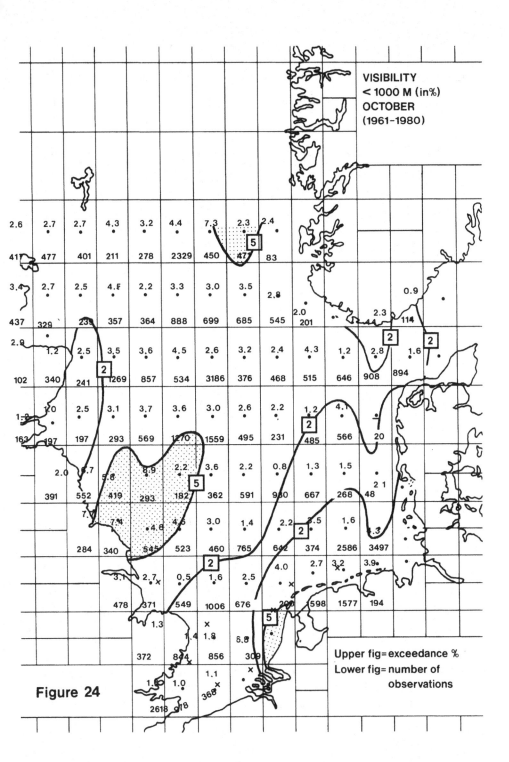

Figure 24

VISIBILITY
< 1000 M (in %)
OCTOBER
(1961–1980)

Upper fig= exceedance %
Lower fig= number of
observations

59

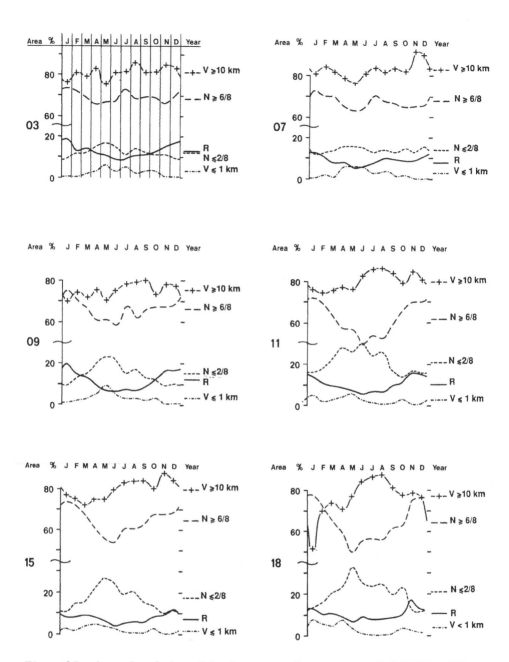

Figure 25a Annual variation of the frequency of occurrence of visibilities (V) <
1 km and ≥ 10 km, cloud cover (N) ≤ 2/8 and ≥ 6/8 and of precipitation (R) for the subareas 03,07,09,11,15 and 18 indicated in figure 1.

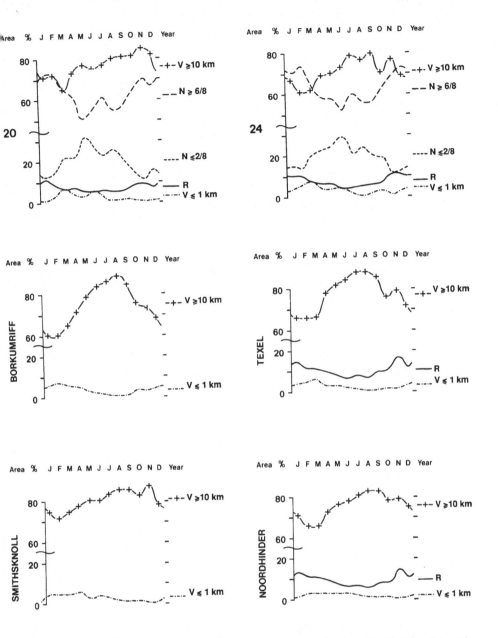

Figure 25b Annual variation of the frequency of occurrence of visibility (V) < 1 km and ≥ 10 km, cloud cover (N) ≤ 2/8 and ≥ 6/8 and of precipitation (R) for the subareas 20 and 24 and for 4 lightvesels indicated in figure 1.

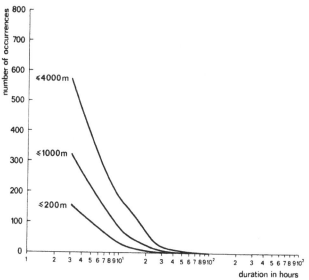

Figure 26   L.V. Texel. Persistence diagram visibility, 1949-1977 (29 years), spring (Mar, Apr, May). Example: Near L.V. Texel in 29 springs 323 periods with visibility ≤ 1000 m occurred, that is on average 11.2 of such periods per spring. Of these 323 periods 178 lasted 6 hours or longer and 16 lasted 24 hours or longer.

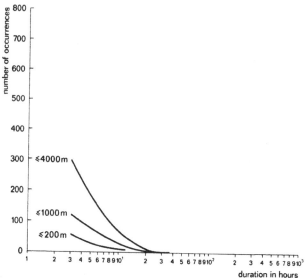

Figure 27   L.V. Texel. Persistence diagram visibility, 1949-1977 ("28 years"), summer (Jun, Jul, Aug). Example: Near L.V. Texel in 28 summers 122 periods with visibility ≤ 1000 m occurred, that is on average 4.4 of such periods per summer. Of these 122 periods 66 lasted 6 hours or longer and 2 lasted 24 hours or longer.

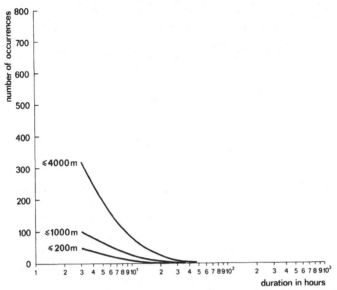

Figure 28 L.V. Texel. Persistence diagram visibility, 1949-1977 ("26 years"),
autumn (Sep, Oct, Nov). Example: Near L.V. Texel in 26 autumns
102 periods with visibility ≤ 1000 m occurred, that is on average 3.9
of such periods per autumn. Of these 102 periods 53 lasted 6 hours or
longer and 3 lasted 24 hours or longer.

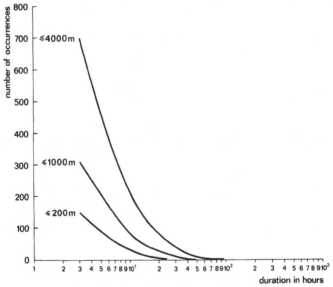

Figure 29 L.V. Texel. Persistence diagram visibility, 1949-1977 (29 years),
winter (Dec, Jan, Feb). Example: Near L.V. Texel in 29 winters 314
periods with visibility ≤ 1000 m occurred, that is on average 10.9 of
such periods per winter. Of these 314 periods 168 lasted 6 hours or
longer and 21 lasted 24 hours or longer.

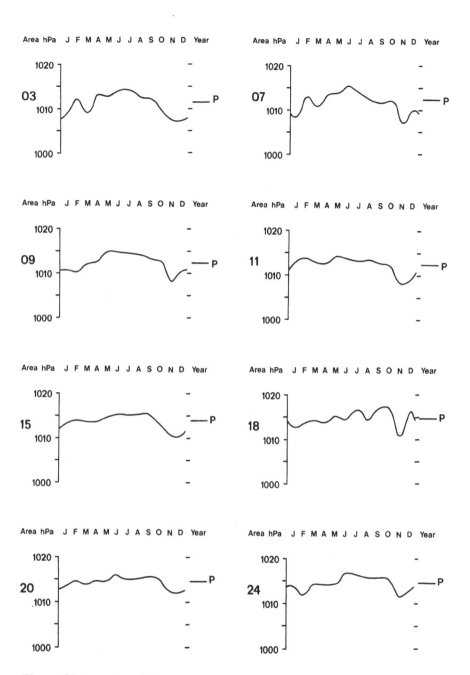

Figure 30 Annual variation of sea level pressure (P) for thesubareas indicated in figure 1.

1971-1980

Area	Jan	Feb	Mar	Apr	May	Jun	Jul	Aug	Sep	Oct	Nov	Dec	Year
03	0.3	0.4	0.3	0.2	0.1	0.2	0.2	0.2	0.4	0.2	0.3	0.3	0.2
07	0.3	0.3	0.3	0.2	0.1	0.2	0.1	0.2	0.3	0.2	0.4	0.3	0.2
09	0.3	0.3	0.2	0.2	0.1	0.4	0.4	0.2	0.3	0.2	0.4	0.4	0.2
11	0.1	0.2	0.2	0.3	0.1	0.5	0.6	0.2	0.4	0.1	0.4	0.2	0.2
15	0.4	0.2	0.2	0.3	0.3	0.3	0.4	0.1	0.4	0.3	0.5	0.4	0.2
18	0.6	0.3	0.0	0.3	0.5	0.3	0.5	0.2	0.3	0.3	0.5	0.5	0.2
20	0.4	0.1	0.1	0.4	0.3	0.3	0.3	0.2	0.4	0.2	0.3	0.5	0.1
24	0.4	0.2	0.3	0.3	0.2	0.3	0.4	0.2	0.4	0.1	0.5	0.4	0.2

1961-1970

Area	Jan	Feb	Mar	Apr	May	Jun	Jul	Aug	Sep	Oct	Nov	Dec	Year
03	0.2	0.2	0.2	0.1	0.1	0.2	0.3	0.2	0.3	0.4	0.1	0.2	0.2
07	0.3	0.1	0.3	0.2	0.0	0.2	0.3	0.2	0.3	0.4	0.2	0.3	0.2
09	0.2	0.1	0.3	0.1	0.1	0.2	0.4	0.3	0.3	0.4	0.1	0.1	0.2
11	0.1	0.1	0.3	0.1	0.1	0.3	0.5	0.3	0.3	0.4	0.0	0.1	0.2
15	0.1	0.1	0.2	0.0	0.1	0.3	0.4	0.3	0.4	0.4	0.2	0.2	0.2
18	0.2	0.1	0.1	0.1	0.0	0.2	0.4	0.2	0.3	0.5	0.3	0.2	0.2
20	0.1	0.2	0.1	0.1	0.1	0.2	0.3	0.4	0.4	0.5	0.2	0.3	0.2
24	0.1	0.2	0.2	0.2	0.3	0.2	0.4	0.3	0.5	0.3	0.2	0.2	0.2

Figures 31,32 Prevailing wind directions and steadiness of the wind for areas indicated in figure 1.

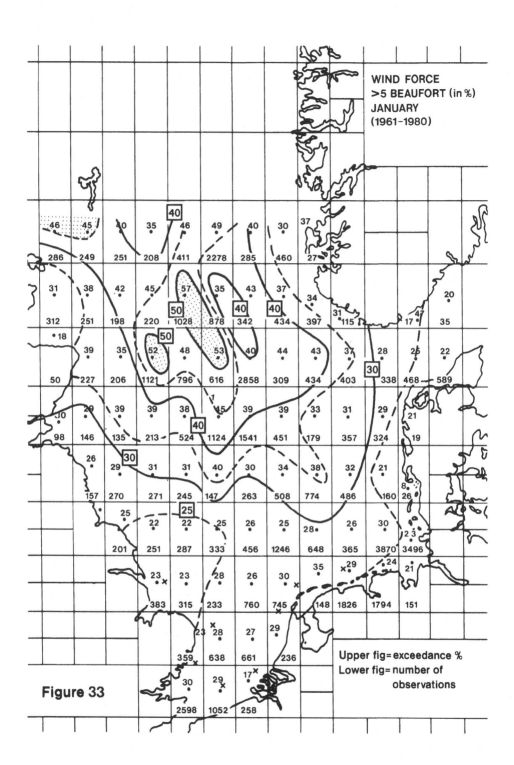

Figure 33

WIND FORCE
>5 BEAUFORT (in %)
JANUARY
(1961-1980)

Upper fig= exceedance %
Lower fig= number of
observations

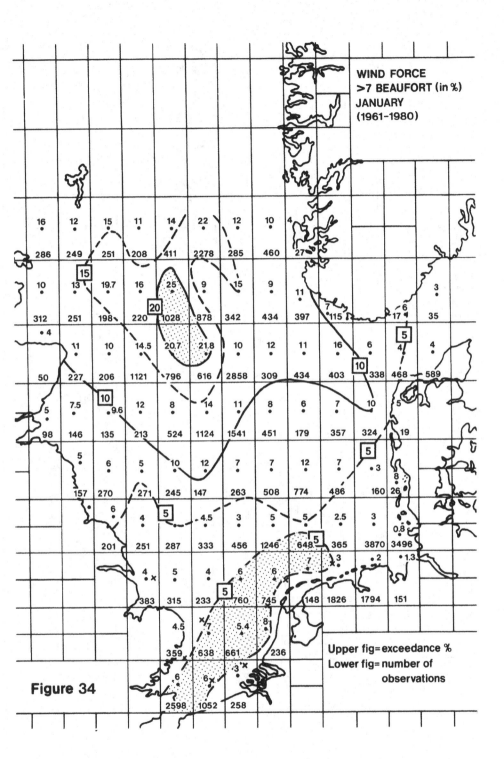

WIND FORCE
>7 BEAUFORT (in %)
JANUARY
(1961-1980)

Upper fig = exceedance %
Lower fig = number of
observations

Figure 34

67

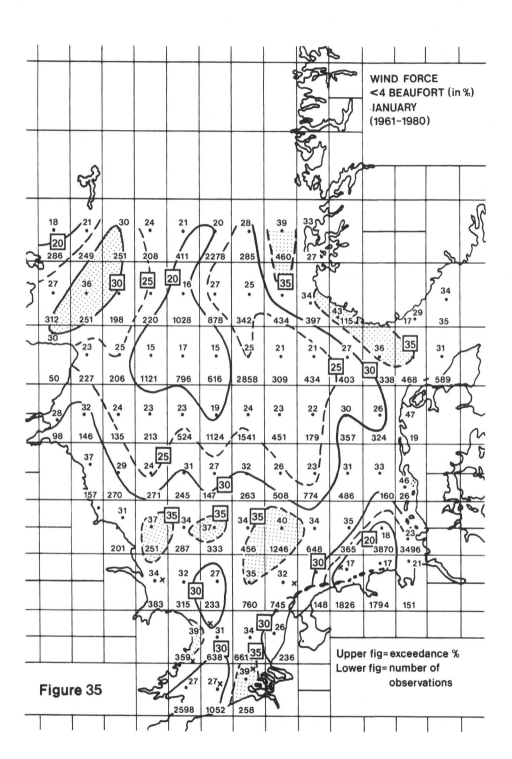

Figure 35

WIND FORCE
<4 BEAUFORT (in %)
JANUARY
(1961-1980)

Upper fig = exceedance %
Lower fig = number of observations

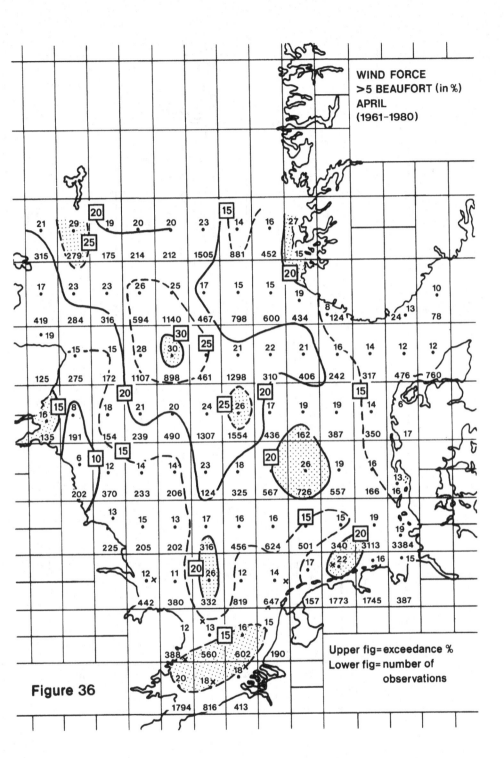

WIND FORCE
>5 BEAUFORT (in %)
APRIL
(1961-1980)

Upper fig= exceedance %
Lower fig= number of observations

Figure 36

69

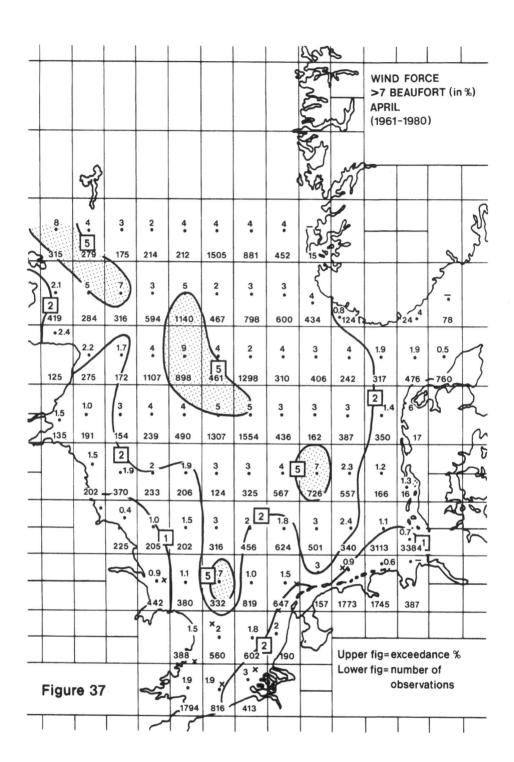

Figure 37

WIND FORCE
>7 BEAUFORT (in %)
APRIL
(1961-1980)

Upper fig= exceedance %
Lower fig= number of
observations

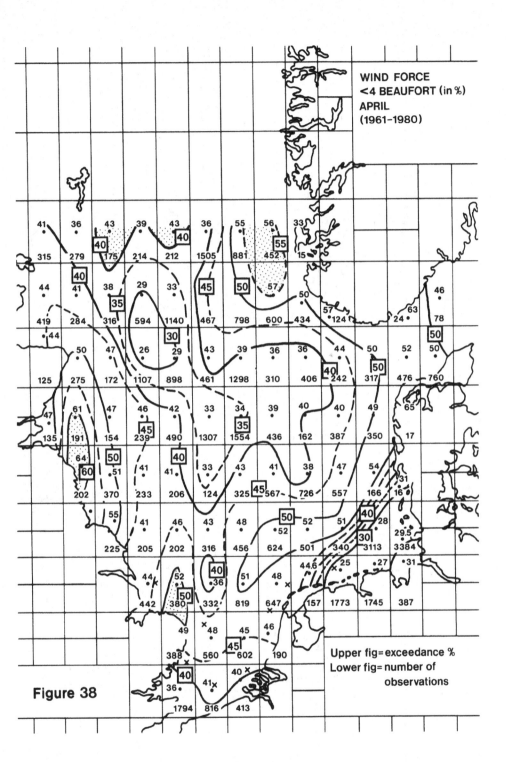

WIND FORCE
<4 BEAUFORT (in %)
APRIL
(1961–1980)

Upper fig=exceedance %
Lower fig= number of
observations

Figure 38

71

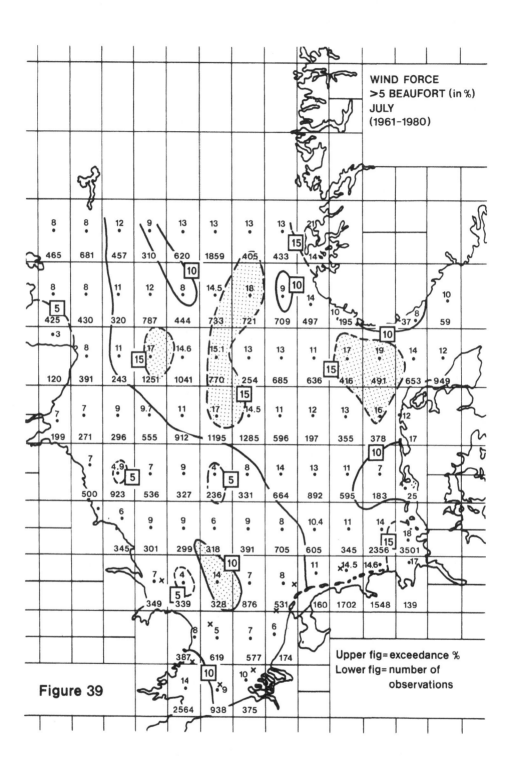

WIND FORCE
>5 BEAUFORT (in %)
JULY
(1961-1980)

Upper fig= exceedance %
Lower fig= number of observations

Figure 39

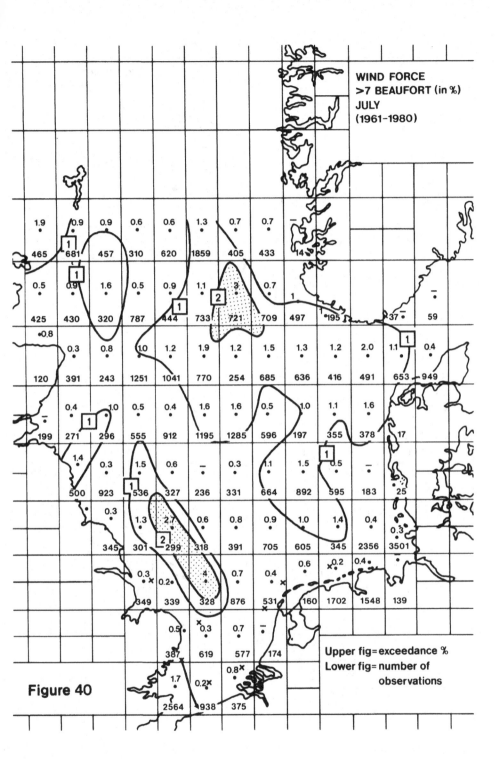

WIND FORCE >7 BEAUFORT (in %) JULY (1961-1980)

Upper fig = exceedance %
Lower fig = number of observations

Figure 40

73

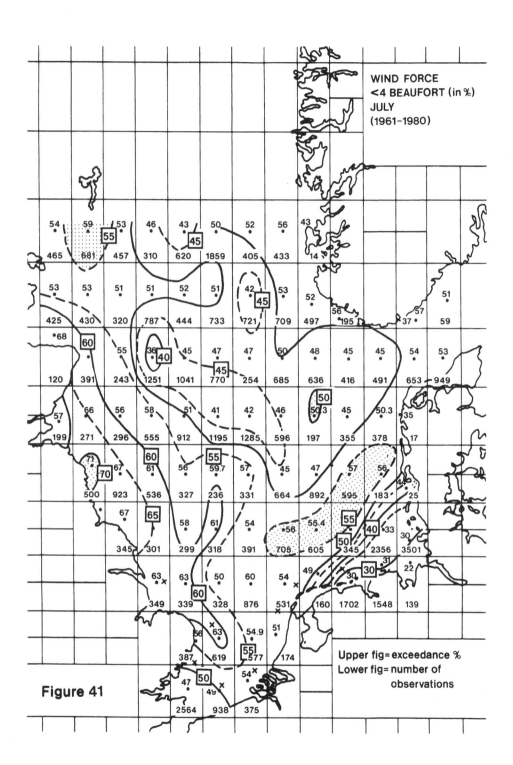

Figure 41

WIND FORCE
<4 BEAUFORT (in %)
JULY
(1961–1980)

Upper fig=exceedance %
Lower fig= number of
observations

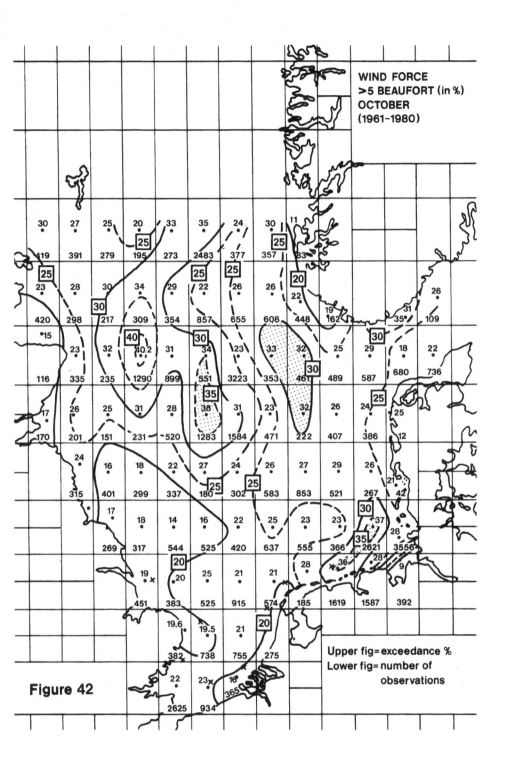

WIND FORCE
>5 BEAUFORT (in %)
OCTOBER
(1961-1980)

Upper fig = exceedance %
Lower fig = number of observations

Figure 42

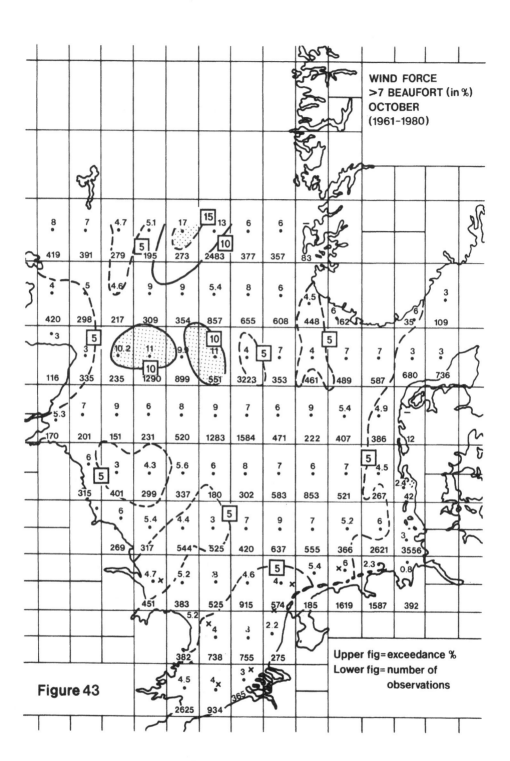

Figure 43

WIND FORCE
>7 BEAUFORT (in %)
OCTOBER
(1961-1980)

Upper fig=exceedance %
Lower fig= number of
observations

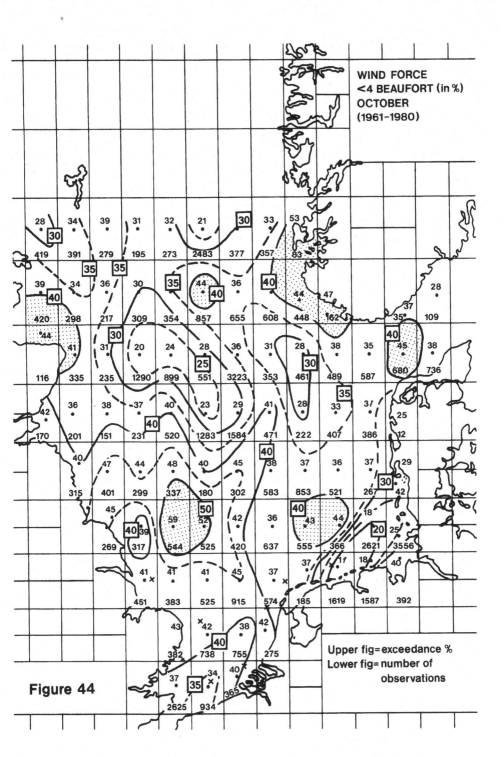

Figure 44

WIND FORCE
<4 BEAUFORT (in %)
OCTOBER
(1961-1980)

Upper fig = exceedance %
Lower fig = number of
observations

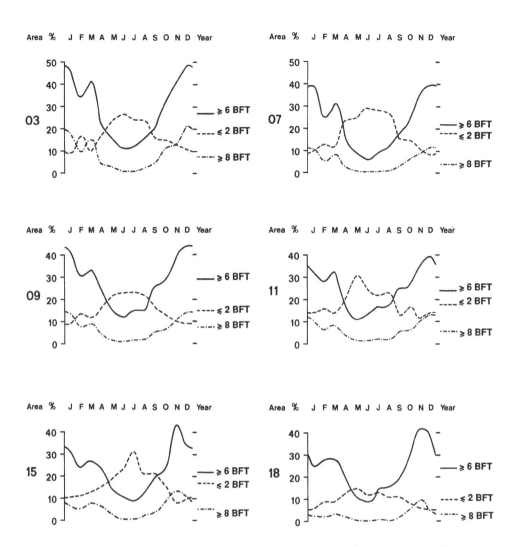

Figure 45a Annual variation of the frequency of occurrence of wind force ≥ 8, ≥ 6 and ≤ 2 for subareas 03,07,09,11,15 and 18 indicated in figure 1.

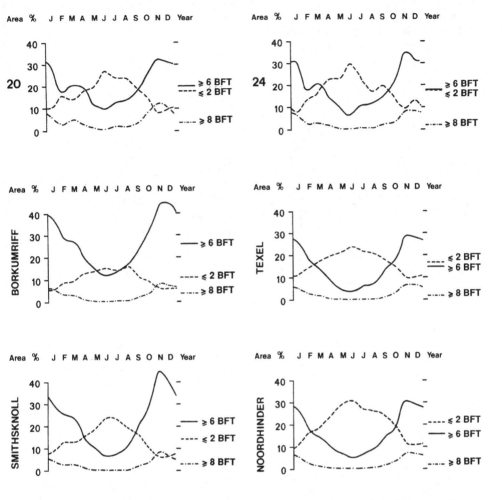

Figure 45b Annual variation of the frequency of occurrence of wind force ≥ 8,
≥ 6 and ≤ 2 for subareas 20 and 24 and 4 lightvessels indicated in
figure 1.

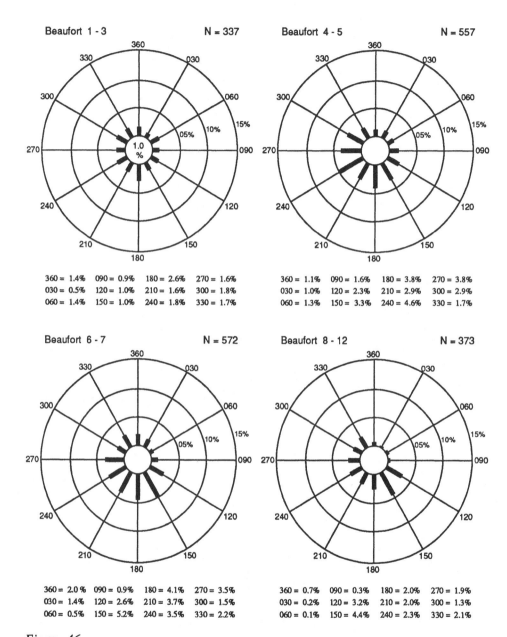

WIND ROSE
January 610101 - 801231 (YYMMDD)
Area: 58.0N - 59.9N, 0.0E - 1.9E N = 1839

Beaufort 1 - 3   N = 337

360 = 1.4%   090 = 0.9%   180 = 2.6%   270 = 1.6%
030 = 0.5%   120 = 1.0%   210 = 1.6%   300 = 1.8%
060 = 1.4%   150 = 1.0%   240 = 1.8%   330 = 1.7%

Beaufort 4 - 5   N = 557

360 = 1.1%   090 = 1.6%   180 = 3.8%   270 = 3.8%
030 = 1.0%   120 = 2.3%   210 = 2.9%   300 = 2.9%
060 = 1.3%   150 = 3.3%   240 = 4.6%   330 = 1.7%

Beaufort 6 - 7   N = 572

360 = 2.0 % 090 = 0.9%   180 = 4.1%   270 = 3.5%
030 = 1.4%   120 = 2.6%   210 = 3.7%   300 = 1.5%
060 = 0.5%   150 = 5.2%   240 = 3.5%   330 = 2.2%

Beaufort 8 - 12   N = 373

360 = 0.7%   090 = 0.3%   180 = 2.0%   270 = 1.9%
030 = 0.2%   120 = 3.2%   210 = 2.0%   300 = 1.3%
060 = 0.1%   150 = 4.4%   240 = 2.3%   330 = 2.1%

Figure 46

80

WIND ROSE
January 610101 - 801231 (YYMMDD)
Area: 54.0N - 55.9N, 2.0E - 3.9E N = 1480

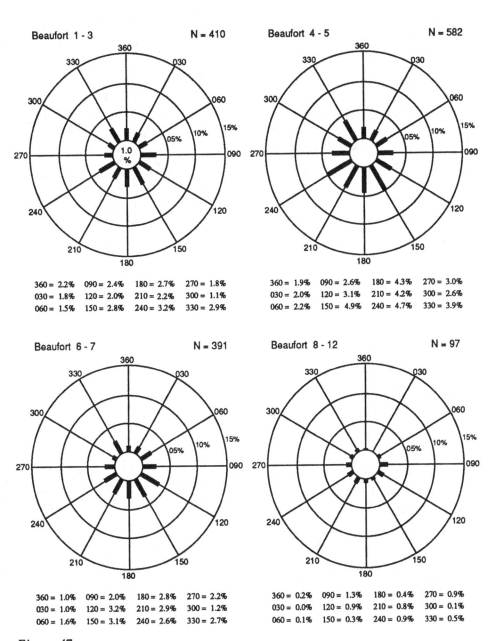

Beaufort 1 - 3   N = 410

360 = 2.2%	090 = 2.4%	180 = 2.7%	270 = 1.8%
030 = 1.8%	120 = 2.0%	210 = 2.2%	300 = 1.1%
060 = 1.5%	150 = 2.8%	240 = 3.2%	330 = 2.9%

Beaufort 4 - 5   N = 582

360 = 1.9%	090 = 2.6%	180 = 4.3%	270 = 3.0%
030 = 2.0%	120 = 3.1%	210 = 4.2%	300 = 2.6%
060 = 2.2%	150 = 4.9%	240 = 4.7%	330 = 3.9%

Beaufort 6 - 7   N = 391

360 = 1.0%	090 = 2.0%	180 = 2.8%	270 = 2.2%
030 = 1.0%	120 = 3.2%	210 = 2.9%	300 = 1.2%
060 = 1.6%	150 = 3.1%	240 = 2.6%	330 = 2.7%

Beaufort 8 - 12   N = 97

360 = 0.2%	090 = 1.3%	180 = 0.4%	270 = 0.9%
030 = 0.0%	120 = 0.9%	210 = 0.8%	300 = 0.1%
060 = 0.1%	150 = 0.3%	240 = 0.9%	330 = 0.5%

Figure 47

81

WIND ROSE
January 610101 - 801231 (YYMMDD)
Area: 52.0N - 52.9N, 3.0E - 4.9E N = 1430

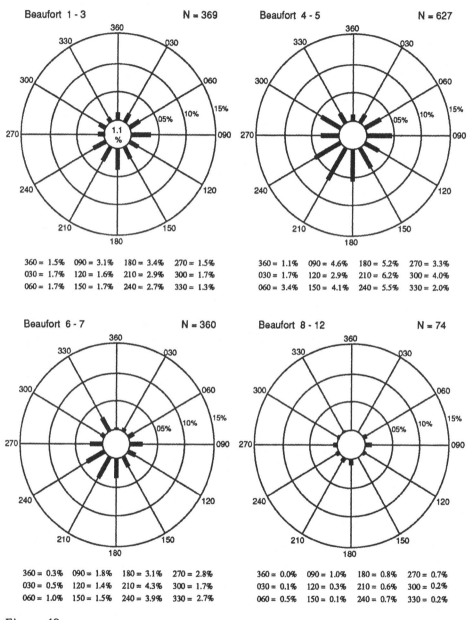

Beaufort 1 - 3 N = 369

360 = 1.5%	090 = 3.1%	180 = 3.4%	270 = 1.5%
030 = 1.7%	120 = 1.6%	210 = 2.9%	300 = 1.7%
060 = 1.7%	150 = 1.7%	240 = 2.7%	330 = 1.3%

Beaufort 4 - 5 N = 627

360 = 1.1%	090 = 4.6%	180 = 5.2%	270 = 3.3%
030 = 1.7%	120 = 2.9%	210 = 6.2%	300 = 4.0%
060 = 3.4%	150 = 4.1%	240 = 5.5%	330 = 2.0%

Beaufort 6 - 7 N = 360

360 = 0.3%	090 = 1.8%	180 = 3.1%	270 = 2.8%
030 = 0.5%	120 = 1.4%	210 = 4.3%	300 = 1.7%
060 = 1.0%	150 = 1.5%	240 = 3.9%	330 = 2.7%

Beaufort 8 - 12 N = 74

360 = 0.0%	090 = 1.0%	180 = 0.8%	270 = 0.7%
030 = 0.1%	120 = 0.3%	210 = 0.6%	300 = 0.2%
060 = 0.5%	150 = 0.1%	240 = 0.7%	330 = 0.2%

Figure 48

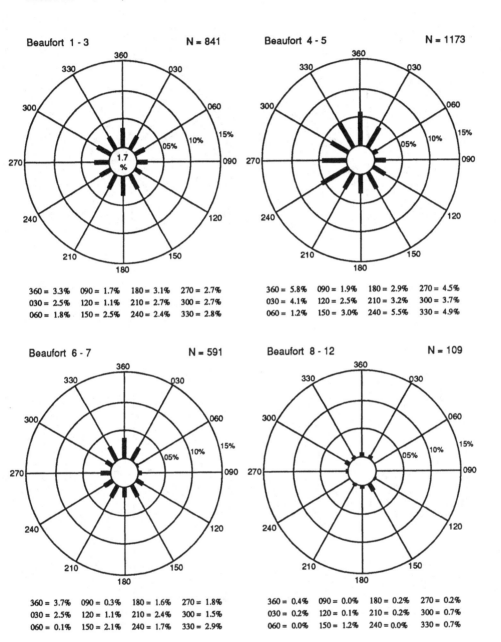

WIND ROSE
April 610101 - 801231 (YYMMDD)
Area: 58.0N - 59.9N, 0.0E - 1.9E N = 2715

Beaufort 1 - 3   N = 841

360 = 3.3%	090 = 1.7%	180 = 3.1%	270 = 2.7%
030 = 2.5%	120 = 1.1%	210 = 2.7%	300 = 2.7%
060 = 1.8%	150 = 2.5%	240 = 2.4%	330 = 2.8%

Beaufort 4 - 5   N = 1173

360 = 5.8%	090 = 1.9%	180 = 2.9%	270 = 4.5%
030 = 4.1%	120 = 2.5%	210 = 3.2%	300 = 3.7%
060 = 1.2%	150 = 3.0%	240 = 5.5%	330 = 4.9%

Beaufort 6 - 7   N = 591

360 = 3.7%	090 = 0.3%	180 = 1.6%	270 = 1.8%
030 = 2.5%	120 = 1.1%	210 = 2.4%	300 = 1.5%
060 = 0.1%	150 = 2.1%	240 = 1.7%	330 = 2.9%

Beaufort 8 - 12   N = 109

360 = 0.4%	090 = 0.0%	180 = 0.2%	270 = 0.2%
030 = 0.2%	120 = 0.1%	210 = 0.2%	300 = 0.7%
060 = 0.0%	150 = 1.2%	240 = 0.0%	330 = 0.7%

Figure 49

WIND ROSE
April 610101 - 801231 (YYMMDD)
Area: 54.0N - 55.9N, 2.0E - 3.9E N = 1458

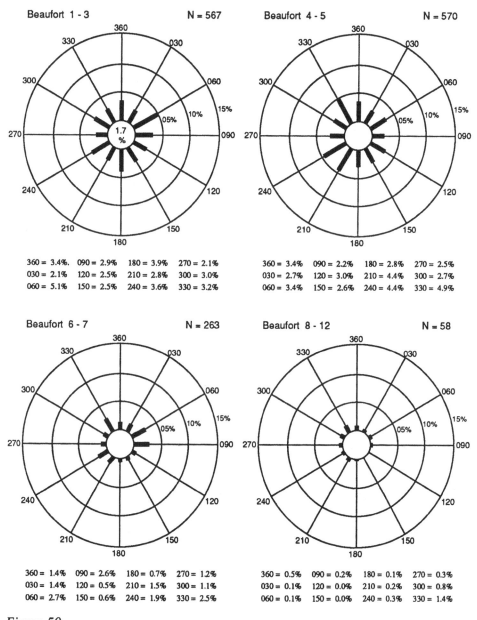

Beaufort 1 - 3   N = 567

360 = 3.4%.   090 = 2.9%   180 = 3.9%   270 = 2.1%
030 = 2.1%   120 = 2.5%   210 = 2.8%   300 = 3.0%
060 = 5.1%   150 = 2.5%   240 = 3.6%   330 = 3.2%

Beaufort 4 - 5   N = 570

360 = 3.4%   090 = 2.2%   180 = 2.8%   270 = 2.5%
030 = 2.7%   120 = 3.0%   210 = 4.4%   300 = 2.7%
060 = 3.4%   150 = 2.6%   240 = 4.4%   330 = 4.9%

Beaufort 6 - 7   N = 263

360 = 1.4%   090 = 2.6%   180 = 0.7%   270 = 1.2%
030 = 1.4%   120 = 0.5%   210 = 1.5%   300 = 1.1%
060 = 2.7%   150 = 0.6%   240 = 1.9%   330 = 2.5%

Beaufort 8 - 12   N = 58

360 = 0.5%   090 = 0.2%   180 = 0.1%   270 = 0.3%
030 = 0.1%   120 = 0.0%   210 = 0.2%   300 = 0.8%
060 = 0.1%   150 = 0.0%   240 = 0.3%   330 = 1.4%

Figure 50

84

WIND ROSE
April 610101 - 801231 (YYMMDD)
Area: 52.0N - 52.9N, 3.0E - 4.9E N = 1262

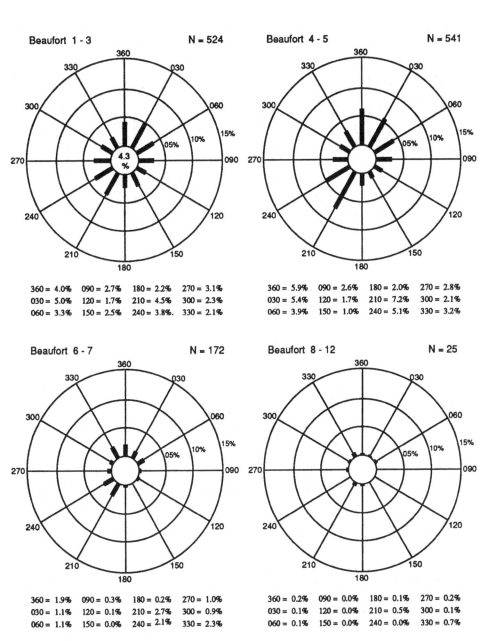

Beaufort 1 - 3   N = 524

360 = 4.0%	090 = 2.7%	180 = 2.2%	270 = 3.1%
030 = 5.0%	120 = 1.7%	210 = 4.5%	300 = 2.3%
060 = 3.3%	150 = 2.5%	240 = 3.8%.	330 = 2.1%

Beaufort 4 - 5   N = 541

360 = 5.9%	090 = 2.6%	180 = 2.0%	270 = 2.8%
030 = 5.4%	120 = 1.7%	210 = 7.2%	300 = 2.1%
060 = 3.9%	150 = 1.0%	240 = 5.1%	330 = 3.2%

Beaufort 6 - 7   N = 172

360 = 1.9%	090 = 0.3%	180 = 0.2%	270 = 1.0%
030 = 1.1%	120 = 0.1%	210 = 2.7%	300 = 0.9%
060 = 1.1%	150 = 0.0%	240 = 2.1%	330 = 2.3%

Beaufort 8 - 12   N = 25

360 = 0.2%	090 = 0.0%	180 = 0.1%	270 = 0.2%
030 = 0.1%	120 = 0.0%	210 = 0.5%	300 = 0.1%
060 = 0.1%	150 = 0.0%	240 = 0.0%	330 = 0.7%

Figure 51

85

WIND ROSE
July 610101 - 801231 (YYMMDD)
Area: 58.0N - 59.9N, 0.0E - 1.9E N = 2434

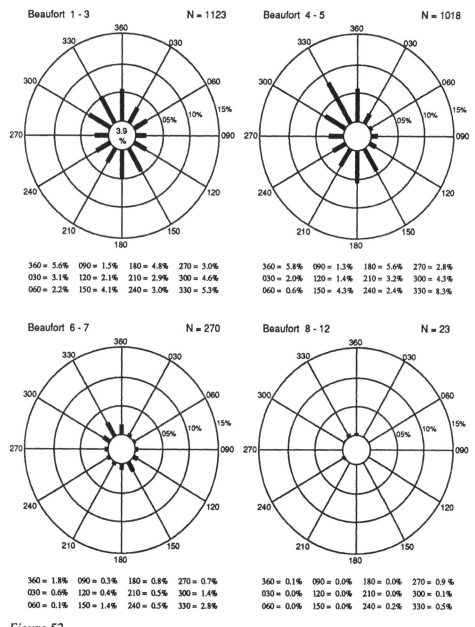

Beaufort 1 - 3   N = 1123

360 = 5.6%	090 = 1.5%	180 = 4.8%	270 = 3.0%
030 = 3.1%	120 = 2.1%	210 = 2.9%	300 = 4.6%
060 = 2.2%	150 = 4.1%	240 = 3.0%	330 = 5.3%

Beaufort 4 - 5   N = 1018

360 = 5.8%	090 = 1.3%	180 = 5.6%	270 = 2.8%
030 = 2.0%	120 = 1.4%	210 = 3.2%	300 = 4.3%
060 = 0.6%	150 = 4.3%	240 = 2.4%	330 = 8.3%

Beaufort 6 - 7   N = 270

360 = 1.8%	090 = 0.3%	180 = 0.8%	270 = 0.7%
030 = 0.6%	120 = 0.4%	210 = 0.5%	300 = 1.4%
060 = 0.1%	150 = 1.4%	240 = 0.5%	330 = 2.8%

Beaufort 8 - 12   N = 23

360 = 0.1%	090 = 0.0%	180 = 0.0%	270 = 0.9 %
030 = 0.0%	120 = 0.0%	210 = 0.0%	300 = 0.1%
060 = 0.0%	150 = 0.0%	240 = 0.2%	330 = 0.5%

Figure 52

WIND ROSE

July 610101 - 801231 (YYMMDD)
Area: 54.0N - 55.9N, 2.0E - 3.9E N = 1484

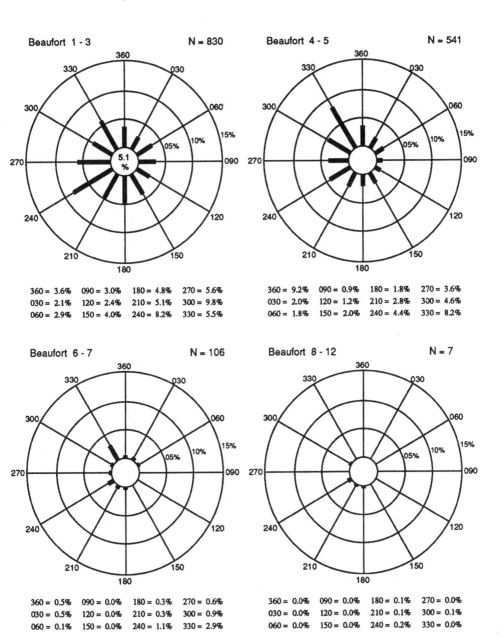

Beaufort 1 - 3   N = 830

360 = 3.6%	090 = 3.0%	180 = 4.8%	270 = 5.6%
030 = 2.1%	120 = 2.4%	210 = 5.1%	300 = 9.8%
060 = 2.9%	150 = 4.0%	240 = 8.2%	330 = 5.5%

Beaufort 4 - 5   N = 541

360 = 9.2%	090 = 0.9%	180 = 1.8%	270 = 3.6%
030 = 2.0%	120 = 1.2%	210 = 2.8%	300 = 4.6%
060 = 1.8%	150 = 2.0%	240 = 4.4%	330 = 8.2%

Beaufort 6 - 7   N = 106

360 = 0.5%	090 = 0.0%	180 = 0.3%	270 = 0.6%
030 = 0.5%	120 = 0.0%	210 = 0.3%	300 = 0.9%
060 = 0.1%	150 = 0.0%	240 = 1.1%	330 = 2.9%

Beaufort 8 - 12   N = 7

360 = 0.0%	090 = 0.0%	180 = 0.1%	270 = 0.0%
030 = 0.0%	120 = 0.0%	210 = 0.1%	300 = 0.1%
060 = 0.0%	150 = 0.0%	240 = 0.2%	330 = 0.0%

Figure 53

87

WIND ROSE

July 610101 - 801231 (YYMMDD)
Area: 52.0N - 52.9N, 3.0E - 4.9E N = 1068

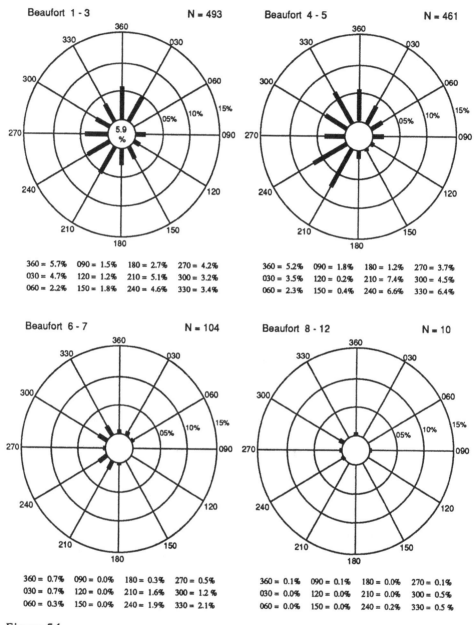

Beaufort 1 - 3 N = 493

360 = 5.7%	090 = 1.5%	180 = 2.7%	270 = 4.2%
030 = 4.7%	120 = 1.2%	210 = 5.1%	300 = 3.2%
060 = 2.2%	150 = 1.8%	240 = 4.6%	330 = 3.4%

Beaufort 4 - 5 N = 461

360 = 5.2%	090 = 1.8%	180 = 1.2%	270 = 3.7%
030 = 3.5%	120 = 0.2%	210 = 7.4%	300 = 4.5%
060 = 2.3%	150 = 0.4%	240 = 6.6%	330 = 6.4%

Beaufort 6 - 7 N = 104

360 = 0.7%	090 = 0.0%	180 = 0.3%	270 = 0.5%
030 = 0.7%	120 = 0.0%	210 = 1.6%	300 = 1.2 %
060 = 0.3%	150 = 0.0%	240 = 1.9%	330 = 2.1%

Beaufort 8 - 12 N = 10

360 = 0.1%	090 = 0.1%	180 = 0.0%	270 = 0.1%
030 = 0.0%	120 = 0.0%	210 = 0.0%	300 = 0.5%
060 = 0.0%	150 = 0.0%	240 = 0.2%	330 = 0.5 %

Figure 54

WIND ROSE

October 610101 - 801231 (YYMMDD)
Area: 58.0N - 59.9N, 0.0E - 1.9E N = 1409

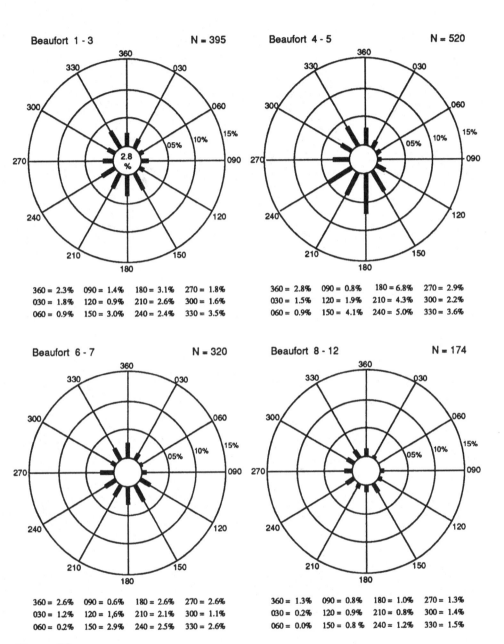

Beaufort 1 - 3   N = 395

360 = 2.3%   090 = 1.4%   180 = 3.1%   270 = 1.8%
030 = 1.8%   120 = 0.9%   210 = 2.6%   300 = 1.6%
060 = 0.9%   150 = 3.0%   240 = 2.4%   330 = 3.5%

Beaufort 4 - 5   N = 520

360 = 2.8%   090 = 0.8%   180 = 6.8%   270 = 2.9%
030 = 1.5%   120 = 1.9%   210 = 4.3%   300 = 2.2%
060 = 0.9%   150 = 4.1%   240 = 5.0%   330 = 3.6%

Beaufort 6 - 7   N = 320

360 = 2.6%   090 = 0.6%   180 = 2.6%   270 = 2.6%
030 = 1.2%   120 = 1.6%   210 = 2.1%   300 = 1.1%
060 = 0.2%   150 = 2.9%   240 = 2.5%   330 = 2.6%

Beaufort 8 - 12   N = 174

360 = 1.3%   090 = 0.8%   180 = 1.0%   270 = 1.3%
030 = 0.2%   120 = 0.9%   210 = 0.8%   300 = 1.4%
060 = 0.0%   150 = 0.8 % 240 = 1.2%   330 = 1.5%

Figure 55

WIND ROSE
October 610101 - 801231 (YYMMDD)
Area: 54.0N - 55.9N, 2.0E - 3.9E N = 1713

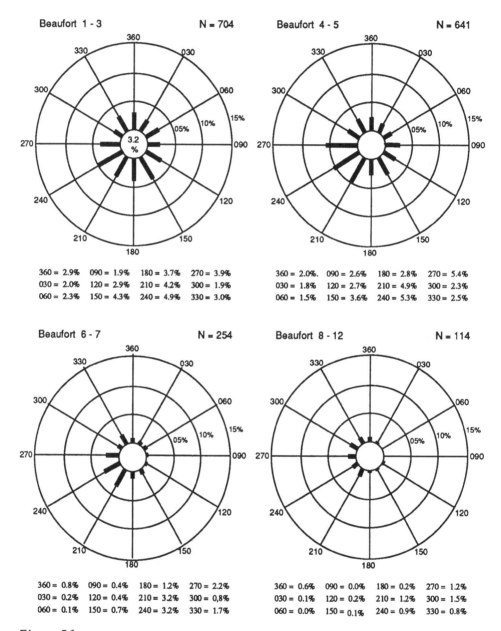

Figure 56

WIND ROSE
October 610101 - 801231 (YYMMDD)
Area: 52.0N - 52.9N, 3.0E - 4.9E N = 1471

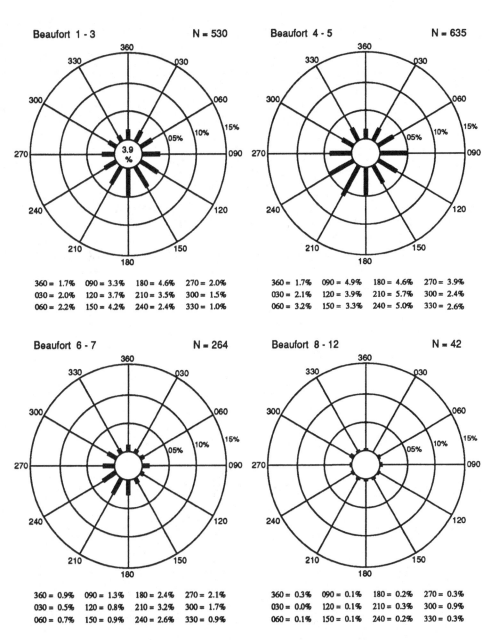

Beaufort 1 - 3   N = 530

360 = 1.7%   090 = 3.3%   180 = 4.6%   270 = 2.0%
030 = 2.0%   120 = 3.7%   210 = 3.5%   300 = 1.5%
060 = 2.2%   150 = 4.2%   240 = 2.4%   330 = 1.0%

Beaufort 4 - 5   N = 635

360 = 1.7%   090 = 4.9%   180 = 4.6%   270 = 3.9%
030 = 2.1%   120 = 3.9%   210 = 5.7%   300 = 2.4%
060 = 3.2%   150 = 3.3%   240 = 5.0%   330 = 2.6%

Beaufort 6 - 7   N = 264

360 = 0.9%   090 = 1.3%   180 = 2.4%   270 = 2.1%
030 = 0.5%   120 = 0.8%   210 = 3.2%   300 = 1.7%
060 = 0.7%   150 = 0.9%   240 = 2.6%   330 = 0.9%

Beaufort 8 - 12   N = 42

360 = 0.3%   090 = 0.1%   180 = 0.2%   270 = 0.3%
030 = 0.0%   120 = 0.1%   210 = 0.3%   300 = 0.9%
060 = 0.1%   150 = 0.1%   240 = 0.2%   330 = 0.3%

Figure 57

91

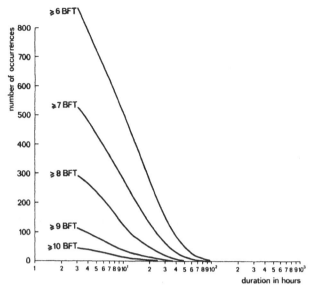

Figure 58   L.V. Texel. Persistence diagram wind, 1949 - 1977 (29 years), winter (Dec, Jan, Feb). Example: Near L.V. Texel in 29 winters 530 periods with wind force 7 or more occurred, that is on average 18.3 of such periods per winter. Of these 530 periods 390 lasted 6 hours or longer and 93 lasted 24 hours or longer.

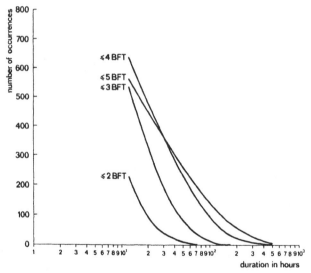

Figure 59   L.V. Texel. Persistence diagram wind, 1949-1977 (29 years), winter (Dec, Jan, Feb). Example: Near L.V. Texel in 29 winters 644 periods with wind force 4 or less lasted 12 hours or longer, that is on average 22.2 of such periods per winter. Of these 644 periods 67 lasted 120 hours (5 days) or longer.

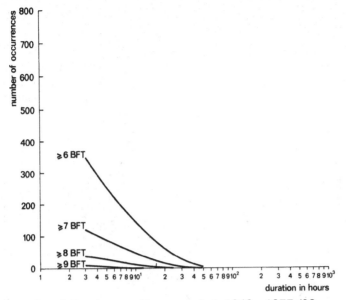

Figure 60   L.V. Texel. Persistence diagram wind, 1949 - 1977 (29 years), summer (Jun, Jul, Aug). Example: Near L.V. Texel in 29 summers 347 periods with force 6 or more occurred, that is on average 12.0 of such periods per summer. Of these 347 periods 219 lasted 6 hours or longer and 45 lasted 24 hours or longer.

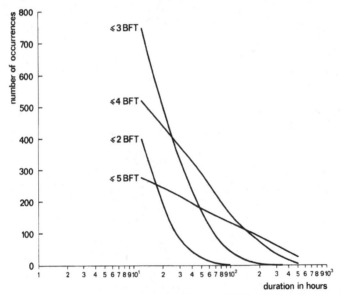

Figure 61   L.V. Texel. Persistence diagram wind, 1949-1977 (29 years), summer (Jun, Jul, Aug). Example: Near L.V. Texel in 29 summers 520 periods with wind force 4 or less lasted 12 hours or longer, that is on average 17.9 of such periods per summer. Of these 520 periods 132 lasted 120 hours (5 days) or longer.

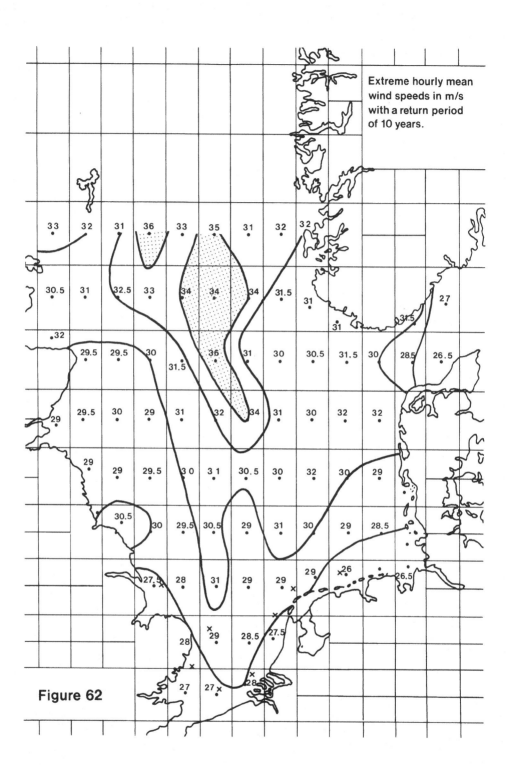

Extreme hourly mean wind speeds in m/s with a return period of 10 years.

Figure 62

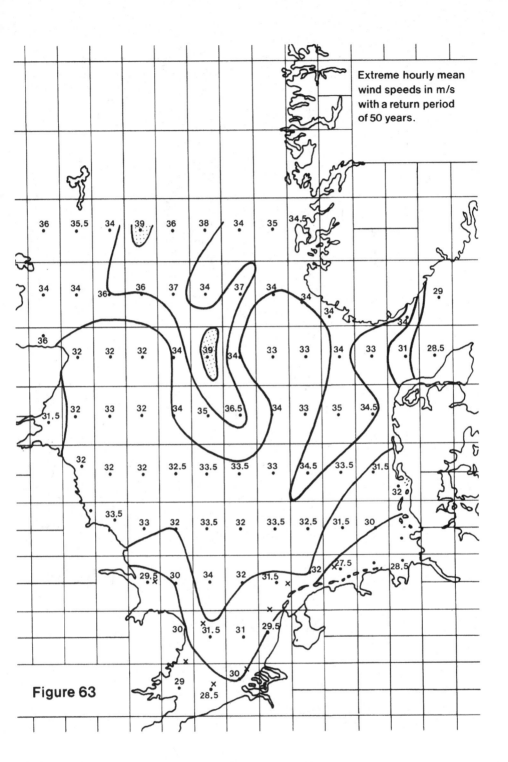

Extreme hourly mean wind speeds in m/s with a return period of 50 years.

Figure 63

95

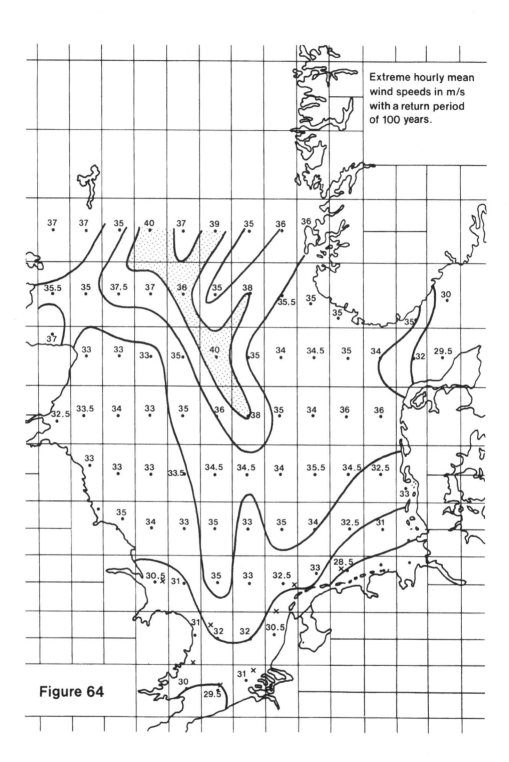

Extreme hourly mean wind speeds in m/s with a return period of 100 years.

Figure 64

Moderate and strong winds, gales and storms (Netherlands lightvessels).
Comparison of frequencies (parts per hundred/fraction of time)

Wind force	Season	Noord Hinder				Schouwen B		Goeree	Haaks	Texel	Terschellingerbank			
		1859 1983	1884 1908	1910 1939	1953 1980	1882 1906	1910 1934	1949 1970	1890 1909	1910 1939	1949 1977	1884 1908	1910 1939	1949 1975
4 and more	Winter	70	60	58	70	39	45	67	41	53	68	47	53	67
	Spring	57	45	37	54	30	32	52	24	33	53	33	35	53
	Summer	51	39	34	47	28	33	52	19	29	48	26	29	47
	Autumn	68	55	51	64	40	44	64	33	48	65	44	49	64
	year	61	50	45	59	34	39	59	30	41	59	38	41	59
6 and more	Winter	34	27	23	25	13	15	24	15	19	24	17	17	22
	Spring	23	16	10	10	7	8	10	6	7	10	9	7	10
	Summer	18	10	7	7	6	7	9	4	5	6	5	5	7
	Autumn	34	23	19	21	13	15	20	12	18	21	17	16	19
	Year	27	19	15	16	10	11	16	9	12	15	12	11	15
8 and more	Winter	9.9	6.7	5.4	4.9	3.1	2.7	4.4	4.3	4.1	4.9	4.3	3.3	3.9
	Spring	4.6	2.4	1.1	0.8	1.1	0.7	0.8	1.1	1.0	1.0	1.1	0.6	1.0
	Summer	2.3	1.1	0.7	0.4	1.0	0.9	0.7	0.6	0.6	0.5	0.5	0.3	0.7
	Autumn	8.5	5.4	4.0	4.2	3.2	2.8	3.5	2.8	3.9	4.2	3.9	3.3	3.0
	Year	6.2	3.8	2.8	2.6	2.1	1.8	2.4	2.2	2.4	2.7	2.4	1.9	2.3
10 and more	Winter	2.0	0.8	0.5	0.4	0.4	0.4	0.4	0.5	0.4	0.5	0.9	0.2	0.5
	Spring	0.7	0.1	0	0	0.1	0.1	0	0.1	0.1	0	0.1	0.1	0
	Summer	0.1	0.1	0	0	0.1	0.1	0	0.1	0	0	-	-	0.1
	Autumn	1.6	0.4	0.3	0.4	0.4	0.3	0.2	0.4	0.6	0.3	0.5	0.5	0.2
	Year	1.0	0.3	0.2	0.2	0.2	0.2	0.2	0.3	0.3	0.2	0.4	0.2	0.2

Figure 65

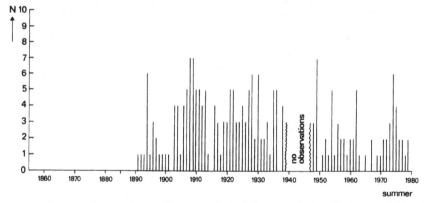

Figure 66   Yearly numbers of storms (wind force ≥ 10) at lightvessel Texel
(L.V. Haaks until 1939) for the years 1891-1980.

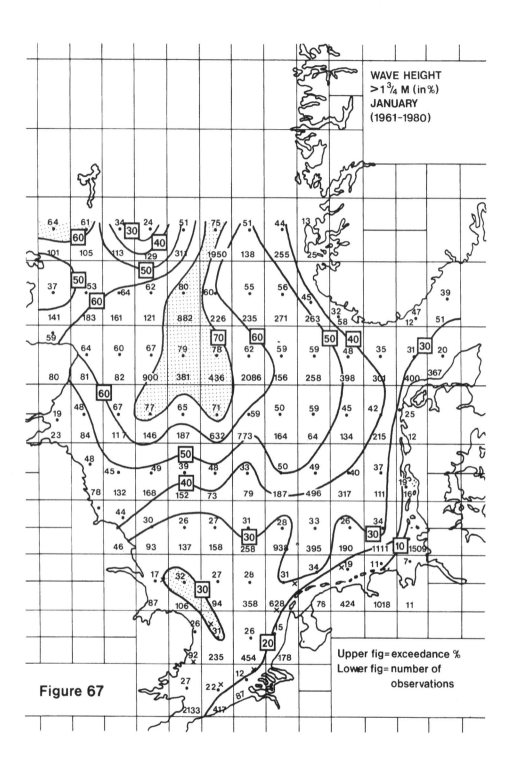

Figure 67

WAVE HEIGHT
$> 1^3/_4$ M (in %)
JANUARY
(1961–1980)

Upper fig = exceedance %
Lower fig = number of
observations

98

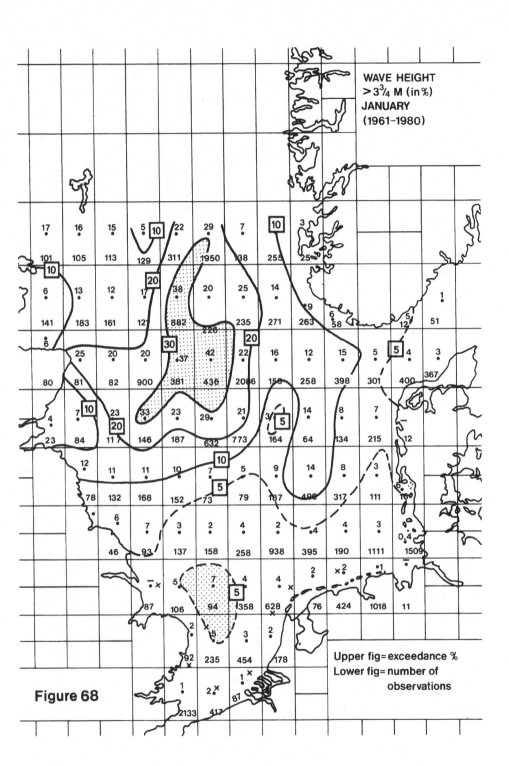

Figure 68

WAVE HEIGHT
$> 3\frac{3}{4}$ M (in %)
JANUARY
(1961–1980)

Upper fig = exceedance %
Lower fig = number of observations

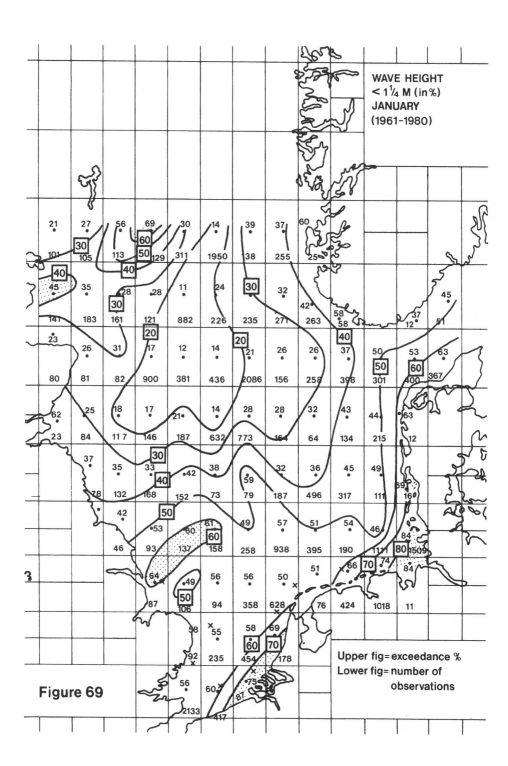

Figure 69

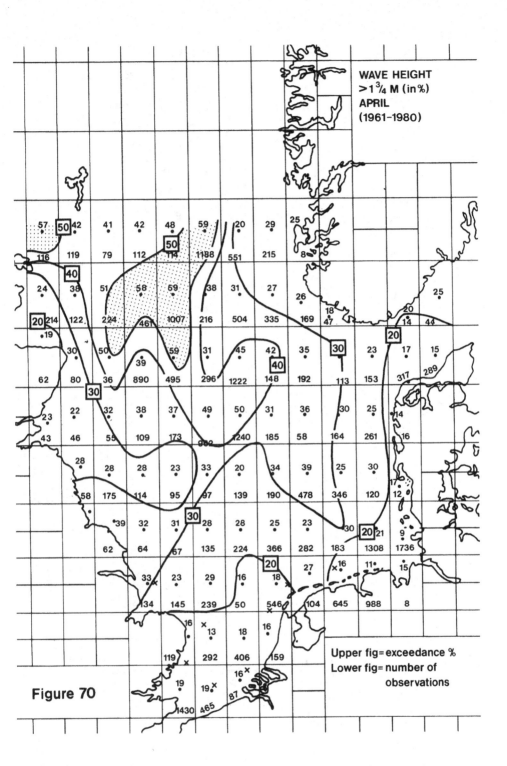

Figure 70

WAVE HEIGHT
$> 1 \frac{3}{4}$ M (in %)
APRIL
(1961–1980)

Upper fig = exceedance %
Lower fig = number of observations

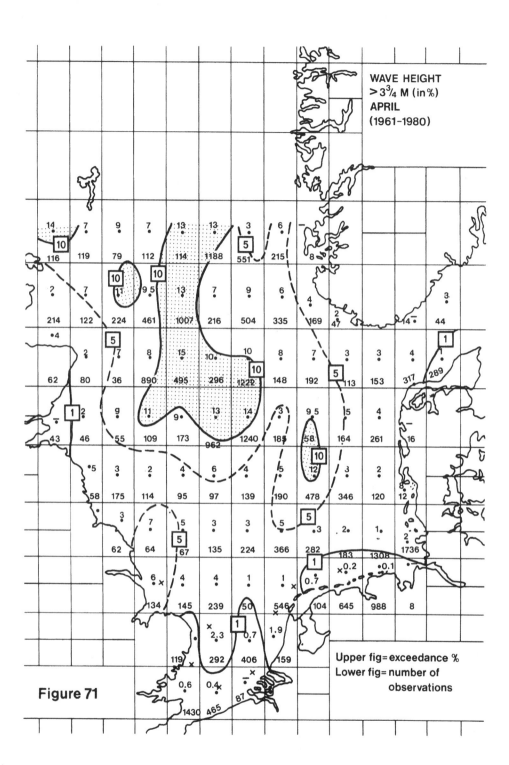

WAVE HEIGHT
> 3¾ M (in %)
APRIL
(1961–1980)

Upper fig = exceedance %
Lower fig = number of
observations

Figure 71

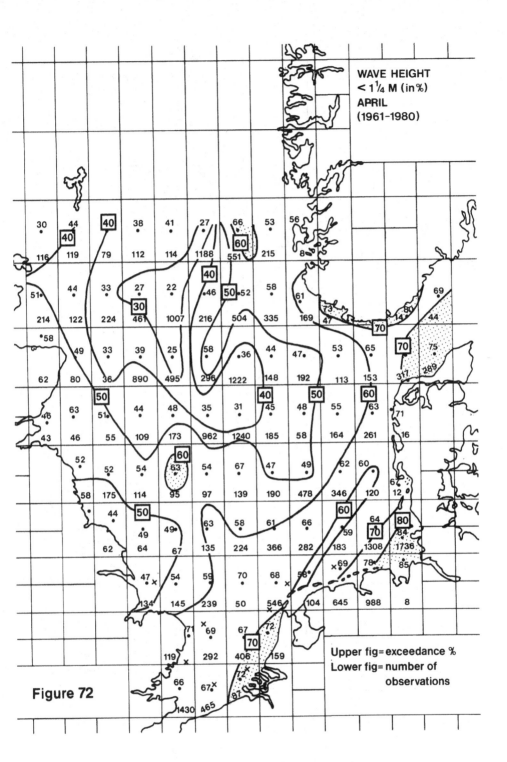

Figure 72

WAVE HEIGHT
< 1¼ M (in %)
APRIL
(1961–1980)

Upper fig = exceedance %
Lower fig = number of observations

103

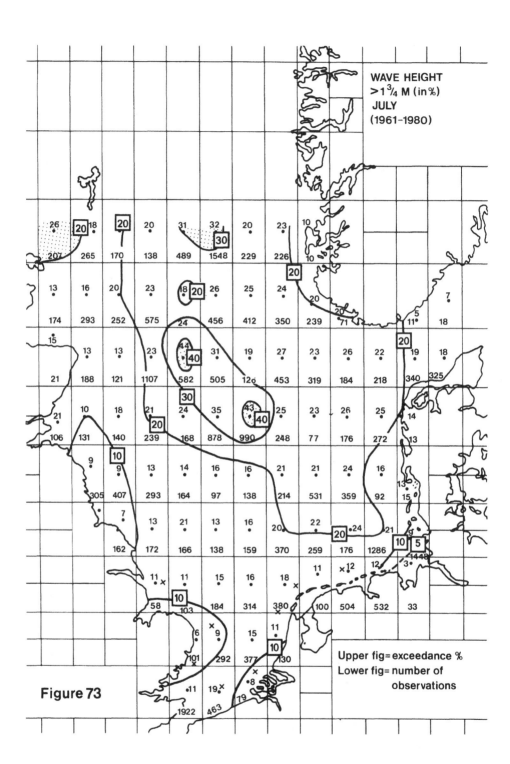

Figure 73

WAVE HEIGHT
>1¾ M (in %)
JULY
(1961–1980)

Upper fig= exceedance %
Lower fig= number of
observations

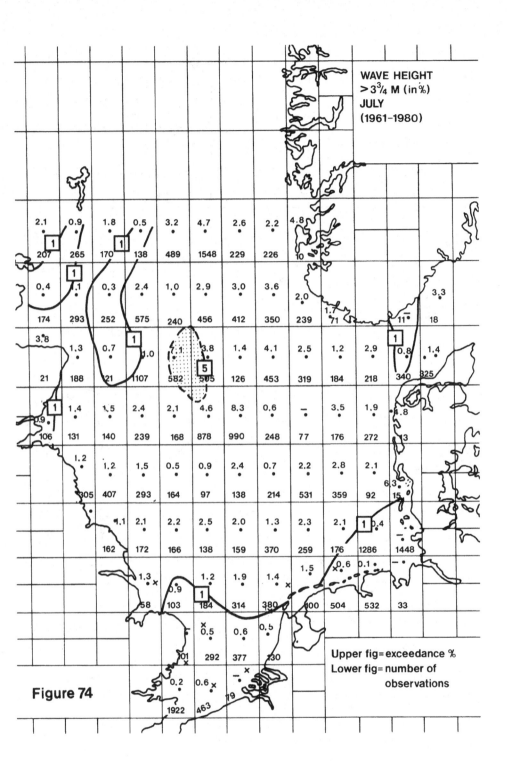

Figure 74

WAVE HEIGHT
$> 3\frac{3}{4}$ M (in %)
JULY
(1961–1980)

Upper fig = exceedance %
Lower fig = number of
                observations

105

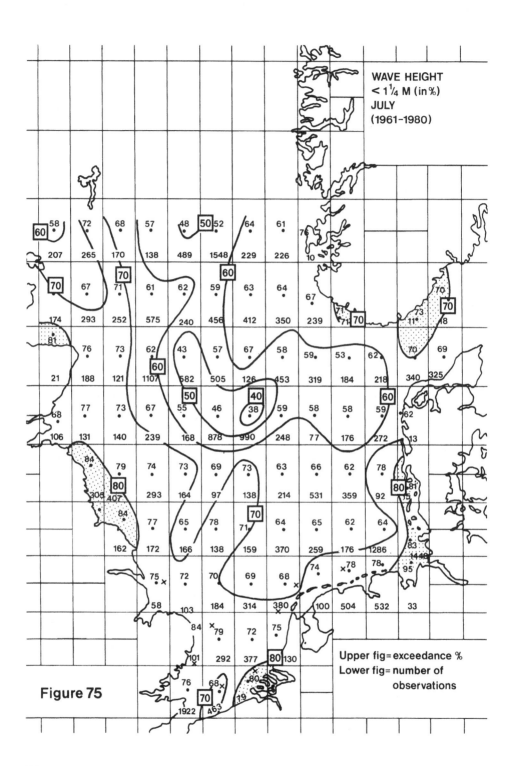

Figure 75

WAVE HEIGHT
< 1¼ M (in %)
JULY
(1961–1980)

Upper fig = exceedance %
Lower fig = number of observations

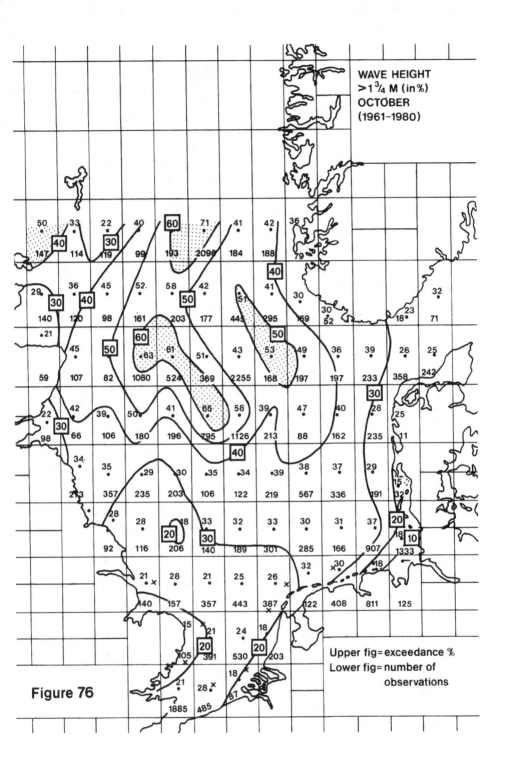

Figure 76

WAVE HEIGHT
>1¾ M (in%)
OCTOBER
(1961–1980)

Upper fig=exceedance %
Lower fig= number of
observations

107

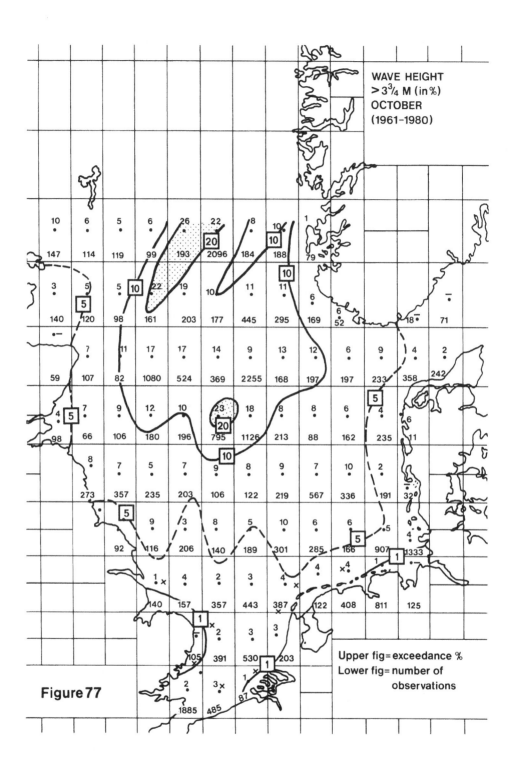

Figure 77

WAVE HEIGHT
> 3¾ M (in %)
OCTOBER
(1961–1980)

Upper fig = exceedance %
Lower fig = number of observations

108

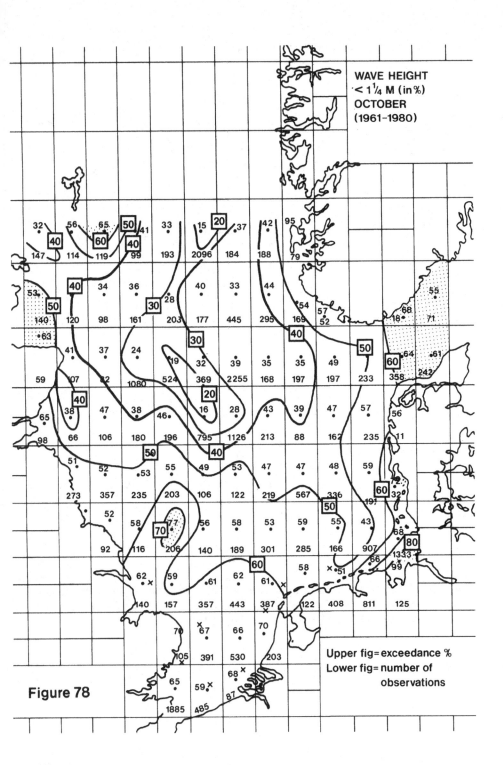

Figure 78

WAVE HEIGHT
< 1¼ M (in %)
OCTOBER
(1961–1980)

Upper fig= exceedance %
Lower fig= number of
observations

109

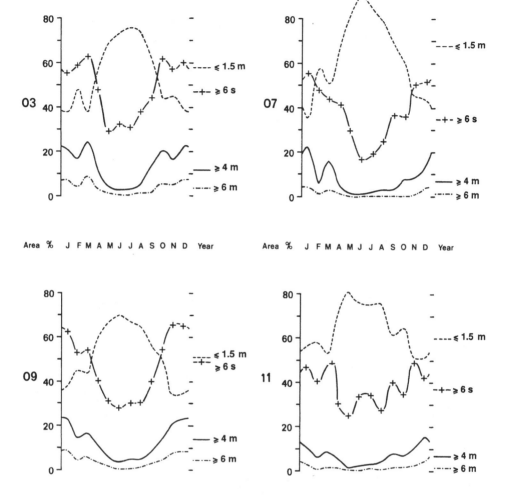

Figure 79a Annual variation of the frequency of occurrence of wave heights ≥ 4m, ≥ 6m, ≤ 1.5m and of wave periods ≥ 6 seconds for subareas 03,07, 09 and 11 indicated in figure 1.

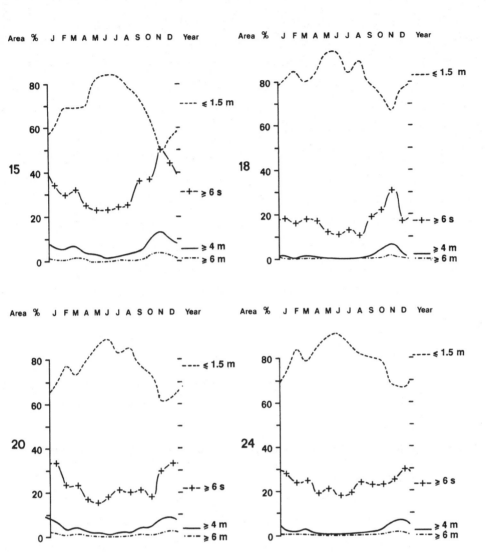

Figure 79b Annual variation of the frequency of occurrence of wave heights ≥ 4m, ≥ 6m, ≤ 1,5m and of wave periods ≥ 6 seconds for subareas 15,18,20 and 24 indicated in figure 1.

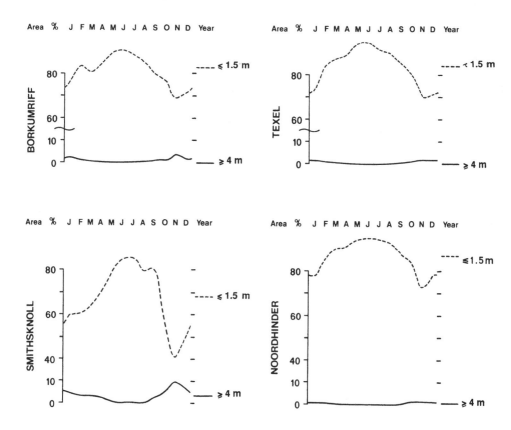

Figure 79c Annual variation of the frequency of occurrence of wave heights ≥ 4m and ≤ 1,5 m at the lightvessels Borkum Riff, Texel, Smiths Knoll and Noord Hinder.

WAVE ROSE
January 1961 - 1980 (wave height in 0.5M values)
Area: 58.0N - 59.9N, 0.0E - 1.9E N = 875

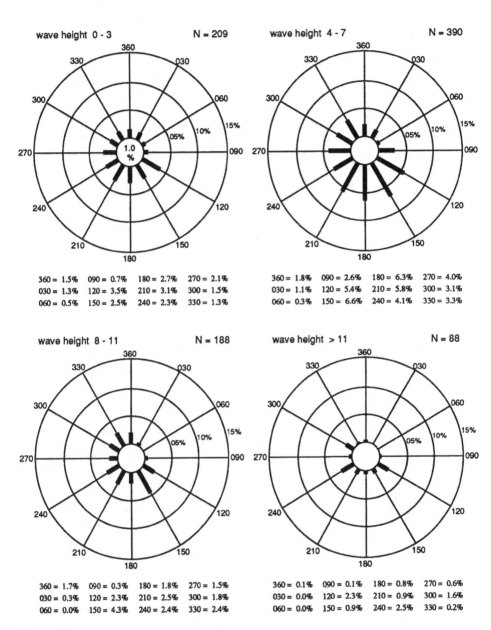

wave height 0 - 3   N = 209

360 = 1.5%	090 = 0.7%	180 = 2.7%	270 = 2.1%
030 = 1.3%	120 = 3.5%	210 = 3.1%	300 = 1.5%
060 = 0.5%	150 = 2.5%	240 = 2.3%	330 = 1.3%

wave height 4 - 7   N = 390

360 = 1.8%	090 = 2.6%	180 = 6.3%	270 = 4.0%
030 = 1.1%	120 = 5.4%	210 = 5.8%	300 = 3.1%
060 = 0.3%	150 = 6.6%	240 = 4.1%	330 = 3.3%

wave height 8 - 11   N = 188

360 = 1.7%	090 = 0.3%	180 = 1.8%	270 = 1.5%
030 = 0.3%	120 = 2.3%	210 = 2.5%	300 = 1.8%
060 = 0.0%	150 = 4.3%	240 = 2.4%	330 = 2.4%

wave height > 11   N = 88

360 = 0.1%	090 = 0.1%	180 = 0.8%	270 = 0.6%
030 = 0.0%	120 = 2.3%	210 = 0.9%	300 = 1.6%
060 = 0.0%	150 = 0.9%	240 = 2.5%	330 = 0.2%

Figuur 80

WAVE ROSE

January 1961 - 1980 (wave height in 0.5M values)
Area: 54.0N - 55.9N, 2.0E - 3.9E N = 808

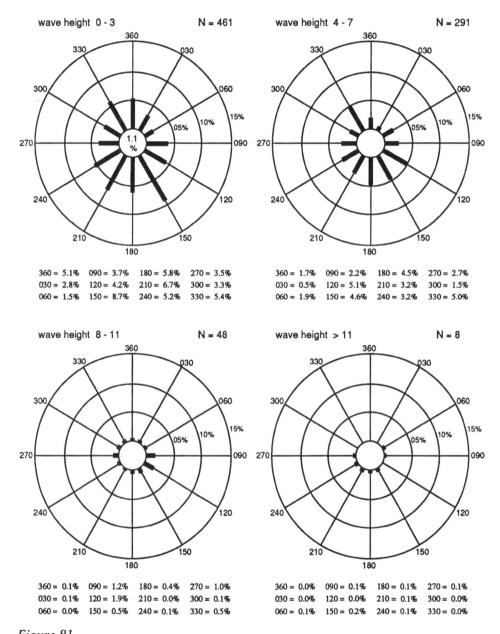

wave height 0 - 3   N = 461

360 = 5.1%	090 = 3.7%	180 = 5.8%	270 = 3.5%
030 = 2.8%	120 = 4.2%	210 = 6.7%	300 = 3.3%
060 = 1.5%	150 = 8.7%	240 = 5.2%	330 = 5.4%

wave height 4 - 7   N = 291

360 = 1.7%	090 = 2.2%	180 = 4.5%	270 = 2.7%
030 = 0.5%	120 = 5.1%	210 = 3.2%	300 = 1.5%
060 = 1.9%	150 = 4.6%	240 = 3.2%	330 = 5.0%

wave height 8 - 11   N = 48

360 = 0.1%	090 = 1.2%	180 = 0.4%	270 = 1.0%
030 = 0.1%	120 = 1.9%	210 = 0.0%	300 = 0.1%
060 = 0.0%	150 = 0.5%	240 = 0.1%	330 = 0.5%

wave height > 11   N = 8

360 = 0.0%	090 = 0.1%	180 = 0.1%	270 = 0.1%
030 = 0.0%	120 = 0.0%	210 = 0.1%	300 = 0.0%
060 = 0.1%	150 = 0.2%	240 = 0.1%	330 = 0.0%

Figure 81

WAVE ROSE

January 1961 - 1980 (wave height in 0.5M values)
Area: 52.0N - 52.9N, 3.0E - 4.9E N = 985

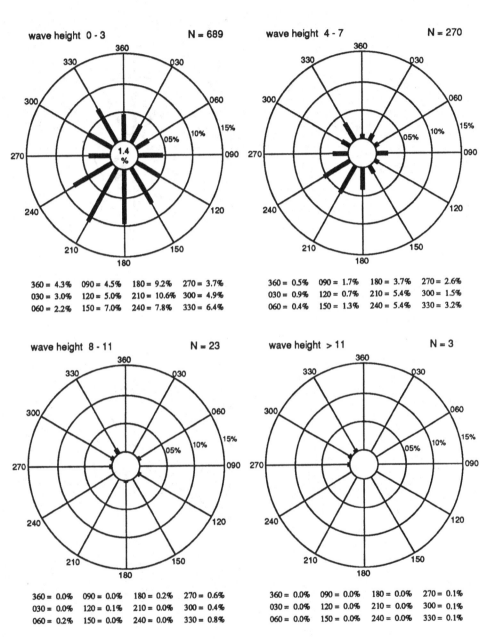

wave height 0 - 3   **N = 689**

360 = 4.3%	090 = 4.5%	180 = 9.2%	270 = 3.7%
030 = 3.0%	120 = 5.0%	210 = 10.6%	300 = 4.9%
060 = 2.2%	150 = 7.0%	240 = 7.8%	330 = 6.4%

wave height 4 - 7   **N = 270**

360 = 0.5%	090 = 1.7%	180 = 3.7%	270 = 2.6%
030 = 0.9%	120 = 0.7%	210 = 5.4%	300 = 1.5%
060 = 0.4%	150 = 1.3%	240 = 5.4%	330 = 3.2%

wave height 8 - 11   **N = 23**

360 = 0.0%	090 = 0.0%	180 = 0.2%	270 = 0.6%
030 = 0.0%	120 = 0.1%	210 = 0.0%	300 = 0.4%
060 = 0.2%	150 = 0.0%	240 = 0.0%	330 = 0.8%

wave height > 11   **N = 3**

360 = 0.0%	090 = 0.0%	180 = 0.0%	270 = 0.1%
030 = 0.0%	120 = 0.0%	210 = 0.0%	300 = 0.1%
060 = 0.0%	150 = 0.0%	240 = 0.0%	330 = 0.1%

Figure 82

115

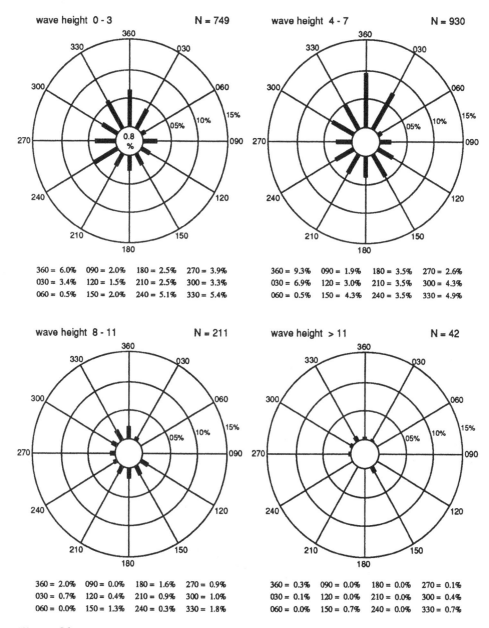

WAVE ROSE
April 1961 - 1980 (wave height in 0.5M values)
Area: 58.0N - 59.9N, 0.0E - 1.9E N = 1932

wave height 0 - 3   N = 749

360 = 6.0%	090 = 2.0%	180 = 2.5%	270 = 3.9%
030 = 3.4%	120 = 1.5%	210 = 2.5%	300 = 3.3%
060 = 0.5%	150 = 2.0%	240 = 5.1%	330 = 5.4%

wave height 4 - 7   N = 930

360 = 9.3%	090 = 1.9%	180 = 3.5%	270 = 2.6%
030 = 6.9%	120 = 3.0%	210 = 3.5%	300 = 4.3%
060 = 0.5%	150 = 4.3%	240 = 3.5%	330 = 4.9%

wave height 8 - 11   N = 211

360 = 2.0%	090 = 0.0%	180 = 1.6%	270 = 0.9%
030 = 0.7%	120 = 0.4%	210 = 0.9%	300 = 1.0%
060 = 0.0%	150 = 1.3%	240 = 0.3%	330 = 1.8%

wave height > 11   N = 42

360 = 0.3%	090 = 0.0%	180 = 0.0%	270 = 0.1%
030 = 0.1%	120 = 0.0%	210 = 0.0%	300 = 0.4%
060 = 0.0%	150 = 0.7%	240 = 0.0%	330 = 0.7%

Figure 83

WAVE ROSE

April 1961 - 1980 (wave height in 0.5M values)
Area: 54.0N - 55.9N, 2.0E - 3.9E N = 781

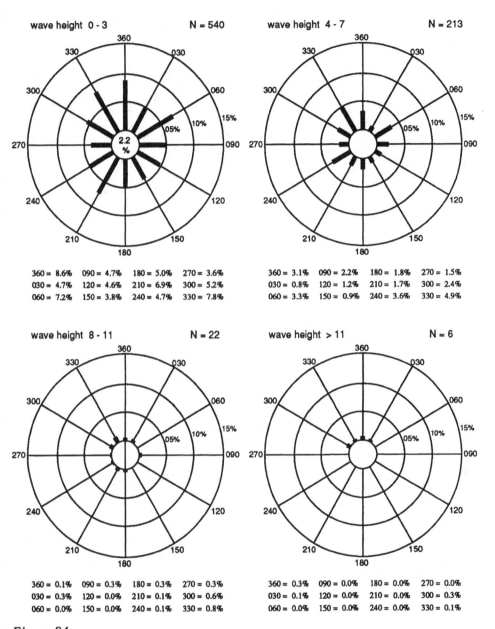

wave height 0 - 3   N = 540

360 = 8.6%   090 = 4.7%   180 = 5.0%   270 = 3.6%
030 = 4.7%   120 = 4.6%   210 = 6.9%   300 = 5.2%
060 = 7.2%   150 = 3.8%   240 = 4.7%   330 = 7.8%

wave height 4 - 7   N = 213

360 = 3.1%   090 = 2.2%   180 = 1.8%   270 = 1.5%
030 = 0.8%   120 = 1.2%   210 = 1.7%   300 = 2.4%
060 = 3.3%   150 = 0.9%   240 = 3.6%   330 = 4.9%

wave height 8 - 11   N = 22

360 = 0.1%   090 = 0.3%   180 = 0.3%   270 = 0.3%
030 = 0.3%   120 = 0.0%   210 = 0.1%   300 = 0.6%
060 = 0.0%   150 = 0.0%   240 = 0.1%   330 = 0.8%

wave height > 11   N = 6

360 = 0.3%   090 = 0.0%   180 = 0.0%   270 = 0.0%
030 = 0.1%   120 = 0.0%   210 = 0.0%   300 = 0.3%
060 = 0.0%   150 = 0.0%   240 = 0.0%   330 = 0.1%

Figure 84

117

WAVE ROSE
April 1961 - 1980 (wave height in 0.5M values)
Area: 52.0N - 52.9N, 3.0E - 4.9E N = 900

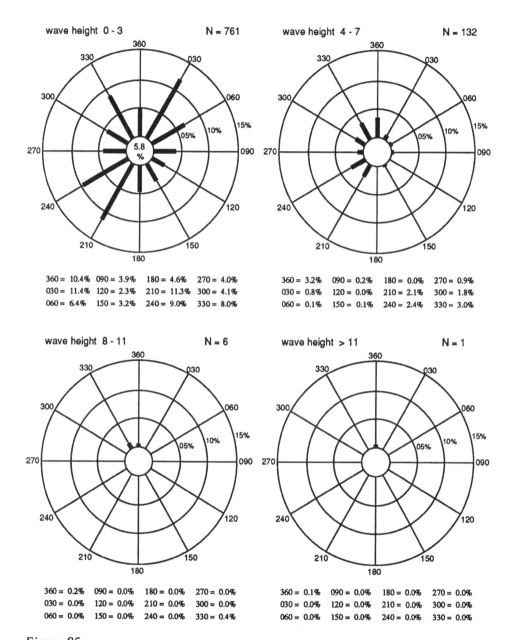

wave height 0 - 3   N = 761

360 = 10.4%	090 = 3.9%	180 = 4.6%	270 = 4.0%
030 = 11.4%	120 = 2.3%	210 = 11.3%	300 = 4.1%
060 = 6.4%	150 = 3.2%	240 = 9.0%	330 = 8.0%

wave height 4 - 7   N = 132

360 = 3.2%	090 = 0.2%	180 = 0.0%	270 = 0.9%
030 = 0.8%	120 = 0.0%	210 = 2.1%	300 = 1.8%
060 = 0.1%	150 = 0.1%	240 = 2.4%	330 = 3.0%

wave height 8 - 11   N = 6

360 = 0.2%	090 = 0.0%	180 = 0.0%	270 = 0.0%
030 = 0.0%	120 = 0.0%	210 = 0.0%	300 = 0.0%
060 = 0.0%	150 = 0.0%	240 = 0.0%	330 = 0.4%

wave height > 11   N = 1

360 = 0.1%	090 = 0.0%	180 = 0.0%	270 = 0.0%
030 = 0.0%	120 = 0.0%	210 = 0.0%	300 = 0.0%
060 = 0.0%	150 = 0.0%	240 = 0.0%	330 = 0.0%

Figure 85

118

WAVE ROSE

July 1961 - 1980 (wave height in 0.5M values)
Area: 58.0N - 59.9N, 0.0E - 1.9E N = 1532

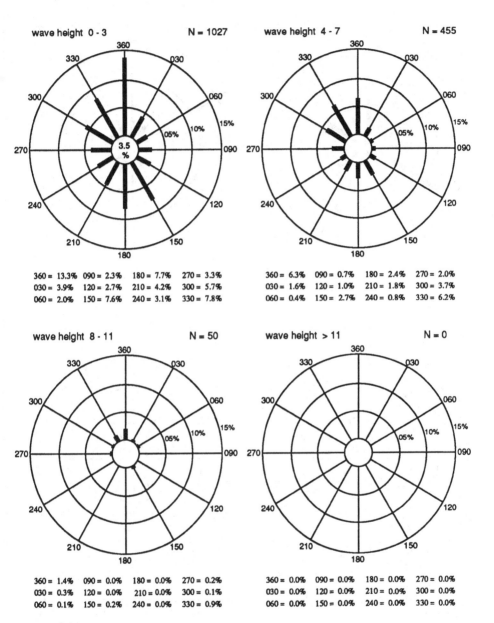

wave height 0 - 3   N = 1027

360 = 13.3%	090 = 2.3%	180 = 7.7%	270 = 3.3%
030 = 3.9%	120 = 2.7%	210 = 4.2%	300 = 5.7%
060 = 2.0%	150 = 7.6%	240 = 3.1%	330 = 7.8%

wave height 4 - 7   N = 455

360 = 6.3%	090 = 0.7%	180 = 2.4%	270 = 2.0%
030 = 1.6%	120 = 1.0%	210 = 1.8%	300 = 3.7%
060 = 0.4%	150 = 2.7%	240 = 0.8%	330 = 6.2%

wave height 8 - 11   N = 50

360 = 1.4%	090 = 0.0%	180 = 0.0%	270 = 0.2%
030 = 0.3%	120 = 0.0%	210 = 0.0%	300 = 0.1%
060 = 0.1%	150 = 0.2%	240 = 0.0%	330 = 0.9%

wave height > 11   N = 0

360 = 0.0%	090 = 0.0%	180 = 0.0%	270 = 0.0%
030 = 0.0%	120 = 0.0%	210 = 0.0%	300 = 0.0%
060 = 0.0%	150 = 0.0%	240 = 0.0%	330 = 0.0%

Figure 86

119

WAVE ROSE
July 1961 - 1980 (wave height in 0.5M values)
Area: 54.0N - 55.9N, 2.0E - 3.9E N = 683

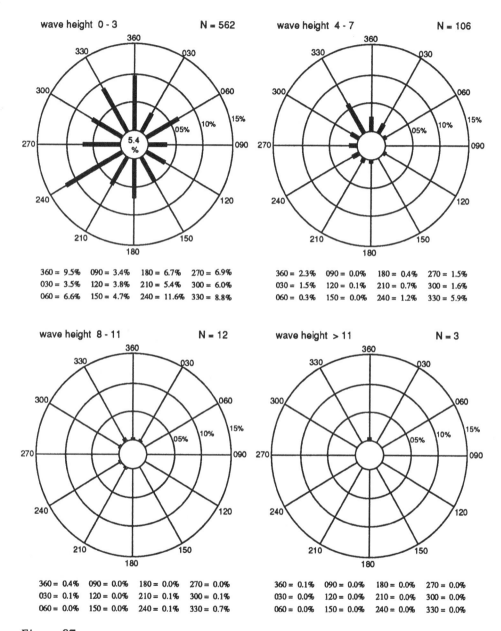

Figure 87

WAVE ROSE

July 1961 - 1980 (wave height in 0.5M values)
Area: 52.0N - 52.9N, 3.0E - 4.9E N = 695

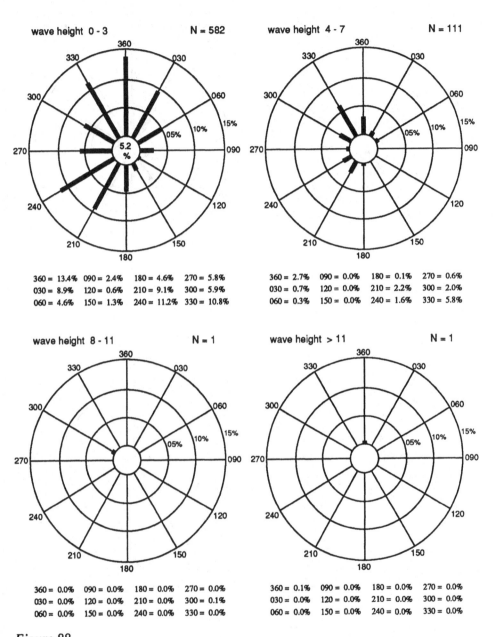

wave height 0 - 3   N = 582

360 = 13.4%	090 = 2.4%	180 = 4.6%	270 = 5.8%
030 = 8.9%	120 = 0.6%	210 = 9.1%	300 = 5.9%
060 = 4.6%	150 = 1.3%	240 = 11.2%	330 = 10.8%

wave height 4 - 7   N = 111

360 = 2.7%	090 = 0.0%	180 = 0.1%	270 = 0.6%
030 = 0.7%	120 = 0.0%	210 = 2.2%	300 = 2.0%
060 = 0.3%	150 = 0.0%	240 = 1.6%	330 = 5.8%

wave height 8 - 11   N = 1

360 = 0.0%	090 = 0.0%	180 = 0.0%	270 = 0.0%
030 = 0.0%	120 = 0.0%	210 = 0.0%	300 = 0.1%
060 = 0.0%	150 = 0.0%	240 = 0.0%	330 = 0.0%

wave height > 11   N = 1

360 = 0.1%	090 = 0.0%	180 = 0.0%	270 = 0.0%
030 = 0.0%	120 = 0.0%	210 = 0.0%	300 = 0.0%
060 = 0.0%	150 = 0.0%	240 = 0.0%	330 = 0.0%

Figure 88

121

WAVE ROSE

October 1961 - 1980 (wave height in 0.5M values)
Area: 58.0N - 59.9N, 0.0E - 1.9E N = 838

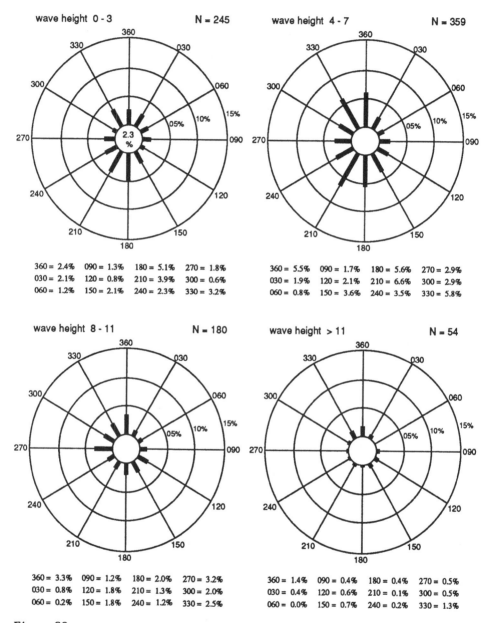

Figure 89

WAVE ROSE

October 1961 - 1980 (wave height in 0.5M values)
Area: 54.0N - 55.9N, 2.0E - 3.9E N = 751

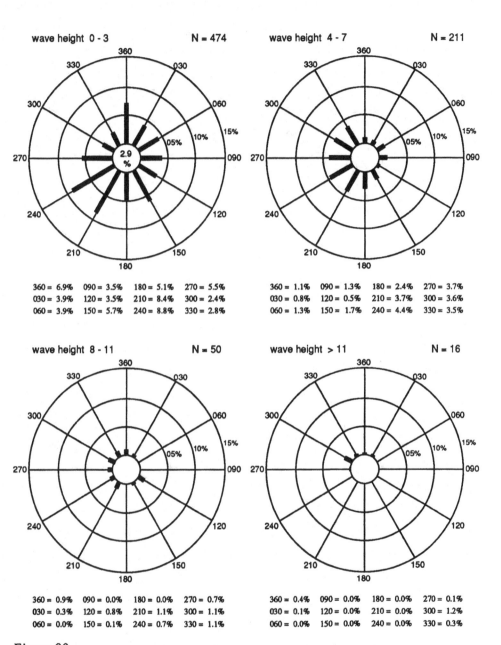

wave height 0 - 3   N = 474

360 = 6.9%   090 = 3.5%   180 = 5.1%   270 = 5.5%
030 = 3.9%   120 = 3.5%   210 = 8.4%   300 = 2.4%
060 = 3.9%   150 = 5.7%   240 = 8.8%   330 = 2.8%

wave height 4 - 7   N = 211

360 = 1.1%   090 = 1.3%   180 = 2.4%   270 = 3.7%
030 = 0.8%   120 = 0.5%   210 = 3.7%   300 = 3.6%
060 = 1.3%   150 = 1.7%   240 = 4.4%   330 = 3.5%

wave height 8 - 11   N = 50

360 = 0.9%   090 = 0.0%   180 = 0.0%   270 = 0.7%
030 = 0.3%   120 = 0.8%   210 = 1.1%   300 = 1.1%
060 = 0.0%   150 = 0.1%   240 = 0.7%   330 = 1.1%

wave height > 11   N = 16

360 = 0.4%   090 = 0.0%   180 = 0.0%   270 = 0.1%
030 = 0.1%   120 = 0.0%   210 = 0.0%   300 = 1.2%
060 = 0.0%   150 = 0.0%   240 = 0.0%   330 = 0.3%

Figure 90

123

WAVE ROSE
October 1961 - 1980 (wave height in 0.5M values)
Area: 52.0N - 52.9N, 3.0E - 4.9E N = 1049

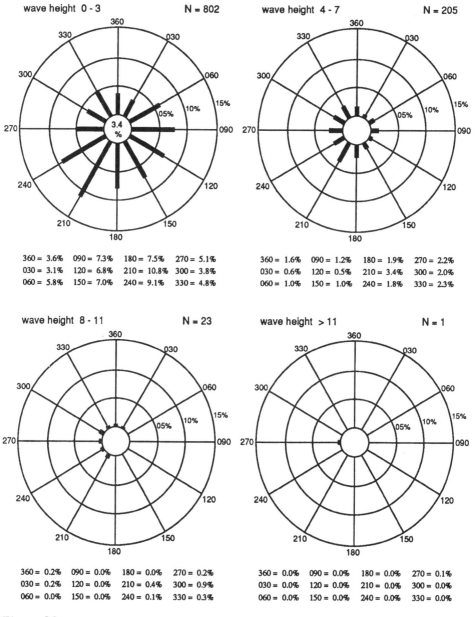

wave height 0 - 3   N = 802

360 = 3.6%	090 = 7.3%	180 = 7.5%	270 = 5.1%
030 = 3.1%	120 = 6.8%	210 = 10.8%	300 = 3.8%
060 = 5.8%	150 = 7.0%	240 = 9.1%	330 = 4.8%

wave height 4 - 7   N = 205

360 = 1.6%	090 = 1.2%	180 = 1.9%	270 = 2.2%
030 = 0.6%	120 = 0.5%	210 = 3.4%	300 = 2.0%
060 = 1.0%	150 = 1.0%	240 = 1.8%	330 = 2.3%

wave height 8 - 11   N = 23

360 = 0.2%	090 = 0.0%	180 = 0.0%	270 = 0.2%
030 = 0.2%	120 = 0.0%	210 = 0.4%	300 = 0.9%
060 = 0.0%	150 = 0.0%	240 = 0.1%	330 = 0.3%

wave height > 11   N = 1

360 = 0.0%	090 = 0.0%	180 = 0.0%	270 = 0.1%
030 = 0.0%	120 = 0.0%	210 = 0.0%	300 = 0.0%
060 = 0.0%	150 = 0.0%	240 = 0.0%	330 = 0.0%

Figure 91

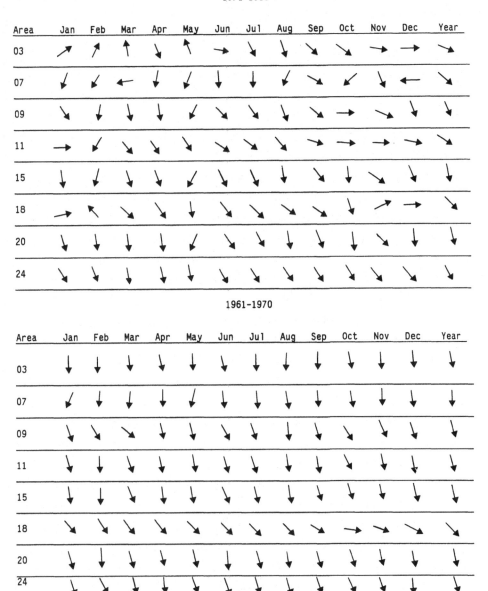

Figure 92 Prevailing swell directions for areas indicated in figure 1.

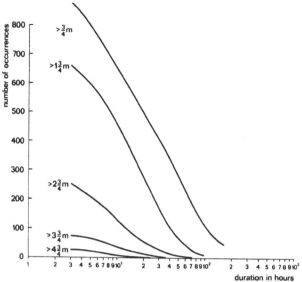

Figure 94   L.V. Texel. Persistence diagram waves, 1949-1977 (29 years), winter (Dec, Jan, Feb). Example: Near L.V. Texel in 29 winters 549 periods, with wave height < 1.25 m lasted 12 hours or longer, that is on average 18.9 of such periods per winter. Of these 549 periods 86 lasted 120 hours (5 days) or longer.

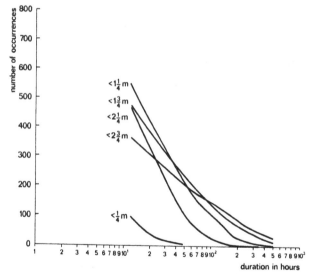

Figure 93   L.V. Texel. Persistence diagram waves, 1949-1977 (29 years), winter (Dec, Jan, Feb). Example: Near L.V. Texel in 29 winters 251 periods with wave height > 2 3/4 m occurred, that is on average 8.7 of such periods per winter. Of these 251 periods 103 lasted 12 hours or longer and 7 lasted 48 hours or longer.

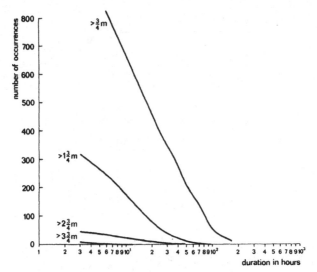

Figure 95   L.V. Texel. Persistence diagram waves, 1949 - 1977 (29 years), summer (Jun, Jul, Aug). Example: Near L.V. Texel in 29 summers 315 periods with wave height > 1 3/4 m occurred, that is on average 10.9 of such periods per summer. Of these 315 periods 150 lasted 12 hours or longer and 13 lasted 48 hours or longer.

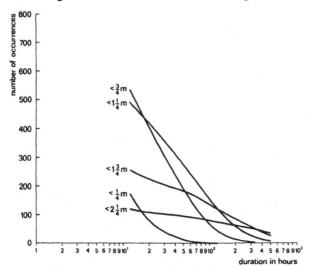

Figure 96   L.V. Texel. Persistence diagram waves, 1949-1977 (29 years), summer (Jun, Jul, Aug). Example: Near L.V. texel in 29 summers 492 periods with wave height < 1.25 m lasted 12 hours or longer, that is on average 17.0 of such periods per summer. Of these 492 periods 121 lasted 120 hours (5 days) or longer and 41 lasted 240 hours (10 days) or longer.

127

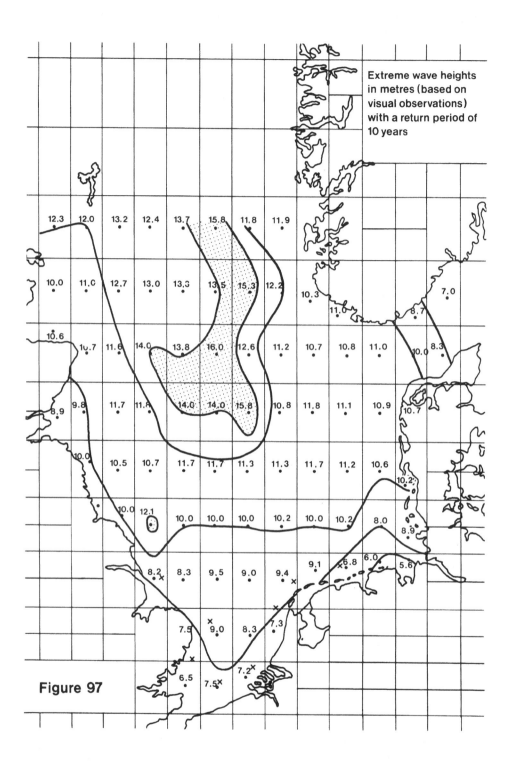

Extreme wave heights in metres (based on visual observations) with a return period of 10 years

Figure 97

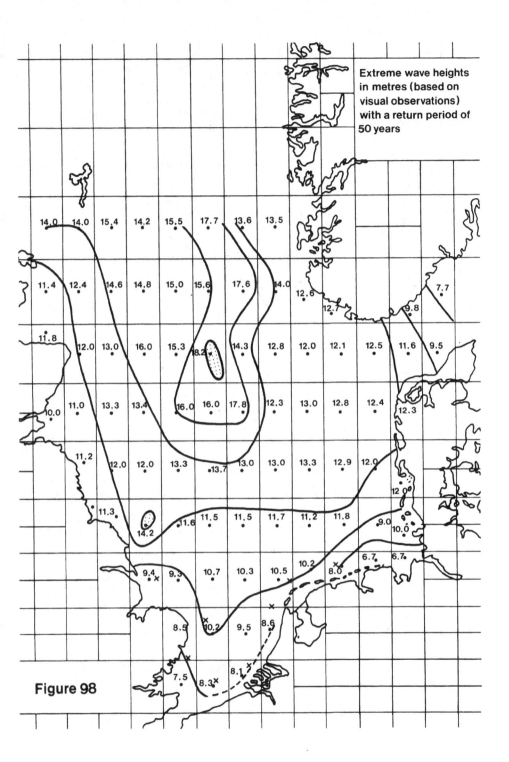

Figure 98

129

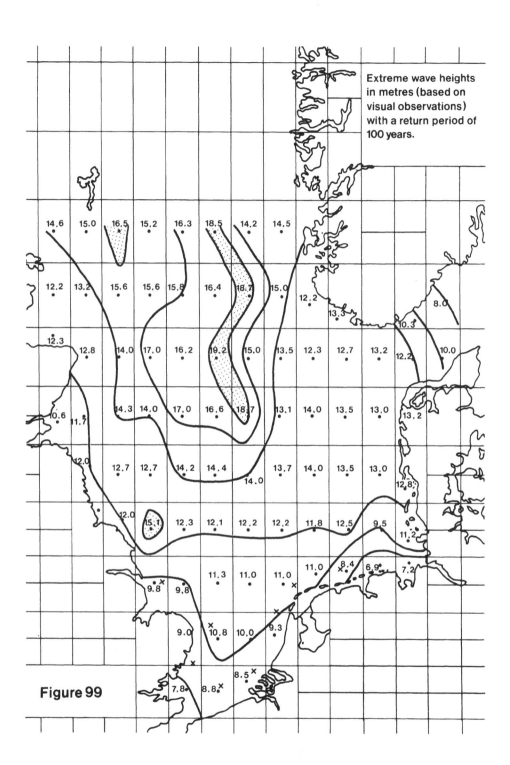

Extreme wave heights in metres (based on visual observations) with a return period of 100 years.

Figure 99

true wind 10

variability of the air temperature 16(11)
variability of the sea surface temperature 17(12)
variation of the mean air temperature 16(13,14)
variation of the mean sea surface temperature 18(15,16)
vector diagram 10
vector mean direction 21,28
vector mean wind speed 21
visibility 3,7,8,18,19(17-29)
visual estimates 3,25,27,29,30
visual observations 7,10,11,27
voluntary observing ships 9,30

water depth 26
water temperature 7
wave crests 10
wave data 3(67-99)
wave height 10,11,26,27,28,29,30(67-79,93-99)
wave period 10,11(79)
wave roses 26,27(80-91)
waves 7,10,26
weather 8
weather charts 3,7
weather observations 3
weather ship data 28
Weibull diagram paper 24,29,30
wind 7,8,21
wind data 3,24(31-66)
wind direction 8,19,21,22,25(31,32,46-57)
wind fetch 9
wind force 9,19,21,22,25(33-45,58-64)
wind force 6 22
wind force 8 22,23
wind force 9 23
wind force 2 23
wind force ≥ 10 25(65,66)
wind roses 22(46-57)
wind speed 9,18,20,21,23,24
wind waves 10
winter 16,17,18,19,22,23,26,28
winter temperatures 16
WMO 3,4,7,9
workable periods 3
World Meteorological organization-WMO 3
World War 16,17,25